Contested Arctic

Indigenous Peoples, Industrial States, and the Circumpolar Environment

Contested Arctic

Indigenous Peoples, Industrial States, and the Circumpolar Environment

Edited by

Eric Alden Smith
Joan McCarter

Preface by Kurt E. Engelmann

Russian, East European, and Central Asian Studies Center
at the Henry M. Jackson School of International Studies
University of Washington

in association with

University of Washington Press
Seattle and London

Printed in the United States of America

Library of Congress Cataloging-in-Publication Data
Contested Arctic: indigenous peoples, industrial states, and the circumpolar environment / edited by Eric Alden Smith, Joan McCarter; preface by Kurt E. Engelmann.
p. cm.
ISBN 0-295-97655-1 (alk. paper)
1. Human ecology—Arctic regions—Congresses. 2. Environmental policy—Arctic regions—Congresses. 3. Arctic peoples—Congresses. 4. Indigenous peoples—Arctic regions—Congresses. 5. Environmental protection—Arctic regions—Congresses. 6. Arctic regions—Environmental conditions—Congresses. I. Smith, Eric Alden. II. McCarter, Joan.
GF891.C66 1997 97-13803
304.2'0911—dc21 CIP

The paper used in this publication meets the minimum requirements of American National Standard for Information Sciences—Permanence of Paper for Printed Library Materials, ANSI Z39.48-1984.

Contents

Maps

Tables

Preface

On May 18, 1996, a one-day symposium on human-environmental interaction in the circumpolar north took place at the University of Washington in Seattle. Co-sponsored by the Russian, East European, and Central Asian Studies (REECAS) Center, in conjunction with the Canadian Studies Center and the Center for West European Studies of the Jackson School of International Studies, the symposium was third in a series of annual symposia on international environmental issues sponsored by the REECAS Center. As in previous years, the symposium was interdisciplinary in nature, with speakers coming from such diverse disciplines as History, Geography, and Anthropology. In contrast with previous years, the symposium featured policy decision-makers along with academic specialists. Papers were divided accordingly, with the first four papers devoted to regional issues of indigenous peoples and environmental problems, and the last two sessions devoted to governmental and intergovernmental policy formation.

Recent political developments in the Russian Federation and elsewhere provided the impetus for the symposium. Contact between indigenous peoples of the north, which has been increasing over the last several decades, accelerated with the opening up of the Soviet Union and the Russian Federation to the outside world. Political openness has led to a vibrant exchange of scholars, activists and political decision-makers at various levels. In the Jackson School, the existence of three area studies centers covering the countries of the circumpolar north provided the institutional framework for arranging the symposium.

The written anthology differs slightly from the papers presented. Molly Fayen, Polar Resource Officer of the U.S. Department of State, gave an official U.S. government account of the Arctic Environmental Protection Strategy at the symposium, which has not been included in the present volume. Fayen's paper has been replaced by a discussion of the Russian Federation approach to pollution problems in the Russian North by Craig ZumBrunnen, who served as a moderator of the symposium.

Several people were instrumental in organizing the symposium. Daniel C. Waugh, Professor of History and former Director of the REECAS Center, originally conceived of the conference and arranged for joint funding from the co-sponsors. Waugh was also important in early discussions between the University of Washington Press and symposium participants. James West, the current REECAS Center Director, has maintained the center's support for the present volume. Susan N. Smith, a recent graduate of the REECAS M.A. Program, helped outline relevant conceptual issues and identified key speakers. Christina Porowski produced maps on West Siberia, Alberta, and indigenous peoples of the circumpolar north.

The late Steven McNabb, an independent anthropologist and expert on Eskimo culture, who died on September 7, 1995 when his canoe capsized in Provideniya Bay in the Russian Far East, contributed his energy and expertise to organizing the symposium, which was dedicated to his memory.

Kurt E. Engelmann
Seattle, Washington
March 25, 1997

Introduction

Eric Alden Smith

Every state contains more than one nation, try as many states might to eliminate or assimilate them all. Every state is multinational; put another way, every state is an empire. And states rule like empires—from the top down.

–Jason Clay (1993)

The Arctic is an early warning system for our planet. . . . There is a link from the rice fields to the ice fields.

–Sergio Marchi, Canadian Minister of Environment (1996)

This book describes a confluence of seemingly disparate events, forces, and cultural realities. It illuminates the intersection of industrial pollution, Arctic ecosystems, national ambitions and indigenous cultures. Its chapters are populated by reindeer herders and caribou hunters, wildlife biologists and oil-rig workers, government bureaucrats and anthropologists. Circumpolar in coverage, the chapters provide detailed accounts of social, political, and environmental interactions or particular locales and peoples; yet they also uncover an underlying parallelism in the growing sociopolitical and ecological problems faced by Arctic cultures.

This brief introduction is meant to highlight these shared themes and parallels. I have grouped my discussion under four headings: (i) the colonial nature of the north's dilemmas; (ii) the connections between ecological problems and indigenous livelihood; (iii) environmental problems as human rights issues; and (iv) emergent solutions.

Northern Colonialism: Business as Usual

In 1976, I went to the Arctic for the first time, to study Inuit subsistence as a graduate student in anthropology. My field site was Inukjuak, a village on the east coast of Hudson Bay in northern Quebec (Smith 1991). My inauguration into fieldwork came just after the First Nations of northern Quebec had signed a land-claims settlement with Canadian federal and provincial governments (DINA 1976). This agreement granted the James Bay Cree of the subarctic forests and the Inuit of the tundra some administrative autonomy and subsistence rights (as well as a multimillion dollar cash settlement) in return for extinguishing aboriginal claims and abandoning their legal challenges to HydroQuebec's massive hydroelectric project in the James Bay region.

Indeed, it was this massive multi-phase hydroelectric project that sparked the politicization of Cree and Inuit. It would eventually divert entire river systems through newly blasted channels, import thousands of Euro-Canadian workers into Cree and Inuit homelands, and install massive turbines and high-voltage electrical lines to turn northern snow-melt into electrical power for homes and factories thousands of miles away (primarily in New York City and New England). Aided by non-Native activists, lawyers, and academics, these northern peoples sued to stop the project and gain legal control over their land, ultimately forcing the out-of-court land-claims settlement. And it was HydroQuebec's project, at that time the largest construction project in the Western Hemisphere, that I glimpsed as I made my way north for a year of fieldwork. I was headed for Great Whale River on the Hudson Bay coast, there to catch a smaller plane headed to Inukjuak. But the old DC-3 that I had flown out of Montréal was full of HydroQuebec workers returning from furlough, so it shuddered to a landing at "LG-2," a staging site blasted and bulldozed out of the north woods. I stared out the window of the plane in the few minutes we sat on the runway, trying to comprehend how this dusty, noisy, scalped and machinery-choked place must appear to the Cree, whose ancestors had hunted, canoed, and camped here for millennia. Though no blood was being spilled or slaves taken away in chains, the Taino "Indians" who watched Columbus sail in, walk ashore, and claim the land for the Spanish Crown could hardly have faced a more alien and destructive presence.

Five centuries after Columbus' landing, the indigenous peoples of the north are experiencing the political and economic expansion of Western states under very different global and local conditions. Yet in a very real sense this is still a colonial expansion. And we must understand this in order to grasp the grave situation faced by contemporary circumpolar peoples, encapsulated as they are within a half-dozen nation-states (expanding empires all, as Jason Clay reminds us). One virtue of this book is the way it drives this message home vividly and repeatedly: the Arctic (along with tropical forests in Amazonia and Southeast Asia) is the frontier of contemporary colonialism (Berger 1993).

This may sound absurd to those accustomed to viewing the present as a postcolonial era. Yet it is no exaggeration to say that Western colonization of the circumpolar north is only now fully underway. Rather than sailing ships, conquistadors, and colonists seeking furs, gold, and farmland, this colonization is carried out by oil-company geologists conducting seismic tests, resource biologists intent on regulating animal populations (as well as the people who depend on these populations for subsistence), and an imported proletariat brought in to work the mines, cut the forests, and dam the rivers. That these contemporary colonial agents often have the best of intentions, or are themselves relatively powerless (as in the case of migrant miners and loggers), indicates that the problem is not a simple one of villainous invaders wantonly destroying lands and peoples out of mere greed. But the complexity of motivation, or the unintended nature of the damage, does not cancel the moral and physical damage wreaked upon the northern colonial frontier.

The resultant social pathology and ecological degradation are usually studied by separate sets of experts—social scientists on the one hand and environmental scientists on the other. But clearly, the social and environmental transformations described in this volume and elsewhere are intimately linked in a single historical process. And since this process is at root a colonial one, it is inevitably political in its causes and hence in its solutions (to the extent solutions can be found). In sum, today's "contested Arctic" is an historical space where three powerful forces collide: cultural, political, and ecological.

Circumpolar Ecosystems and Indigenous Livelihood

The ecological dimension of northern colonialism is prominent for several reasons. First, the colonial impetus in the Arctic is resource extraction rather than settlement or labor control. This extraction—of fossil fuels, timber, minerals, or hydroelectric power—is often pursued in the quickest but most damaging manner. As detailed by Espiritu (chapter 3), Fondahl (chapter 4), and ZumBrunnen (chapter 5), various Native peoples of the Russian and Canadian north share the common fate of seeing their homelands polluted by industrial effluent and oil pipeline leaks, scarred by mining and seismic testing, clear-cut of timber, and overhunted by migrant laborers seeking to supplement their wages with whatever game they can bag. Short-term gains are thus purchased at the cost of severe long-term damage to northern biotic systems, which because of their low species diversity and low productivity are both easily damaged and very slow to repair.

Second, even well-intentioned agents of colonial development are generally unfamiliar with the structure and dynamics of northern ecosystems. As a result, they are more likely to pursue policies and practices that have ecologically (and often economically) damaging unforeseen consequences. This is compounded by the fact that, from the temperate-zone perspective that prevails among colonial and industrial decision-makers in the Arctic, the northern frontier seems vast and underpopulated, and hence purportedly able to absorb the damage caused by extractive activities. Moreover, this frontier is generally distant from the observing eye of mass media, so negative publicity concerning ecological damage is relatively less likely.

Third, ecological degradation in the Arctic also results from industrial and military activities that occur far outside the region. As Charles Johnson details (chapter 1), this "transboundary pollution" of persistent organic pollutants, heavy metals, and other toxic products of industrial production is a growing threat to Arctic peoples. This threat was also a theme emphasized by Sergio Marchi, Canadian Minister of Environment, in a 1996 talk he gave in Ottawa to the Arctic Council, a newly formed advisory body with representatives from all the circumpolar nations. As Marchi pointed out, there is now clear evidence that pesticides sprayed in rice fields in Louisiana are carried by wind and rain to the food chains of the Arctic, and PCBs or heavy metals produced in industrial "southern" cities such as Cleveland or Hamburg or Winnipeg end up in the organs of polar bears and the umbilical cords of newborn Inuit (Schneider 1996). This echoes events of a generation ago, when atmospheric nuclear testing in lower latitudes (including on Pacific atolls whose inhabitants were ordered to leave so their home could be rendered indefinitely uninhabitable in the interests of their colonial masters) resulted in radioactive substances in the breast milk of Arctic mothers. None of these outcomes were intended or in most cases foreseen. They result from atmospheric and marine circulation of contaminants, which become concentrated as they move up Arctic food chains and enter the bodies of Arctic peoples, whose traditional subsistence generally makes them top carnivores in these food chains. Correct as this empirical accounting may be, and innocent of malevolent design though the industrial powers at fault may be, the damage they inflict on Arctic peoples is severe nonetheless.

Finally, the ecological dimension to northern colonialism is of paramount significance because of the ways in which colonial intrusion has so regularly and repeatedly disrupted the subsistence regimes of indigenous peoples. As has been true throughout the colonial expansion of Europeans and their descendants, and is now true for industrial expansion whatever the cultural or ethnic background of the dominant party, the most far-reaching and irreversible imperialism has been ecological (Crosby 1986). Hence we cannot consider environmental problems without considering their sociopolitical causes and consequences.

Environmental Problems and Human Rights

In the circumpolar north, indigenous peoples whose varied traditions have been evolving for millennia increasingly find their cultural and even physical survival hanging in the balance. As documented by various authors herein, healthy indigenous communities are simply not compatible with industrial development or large-scale commercial resource extraction. The linchpin in this conflict is very often the indigenous subsistence system, which is vulnerable to loss of land, loss of local resource management, and, in the worst case, loss of the local resource populations themselves. In any case, once one examines the sociopolitical context, it become clear that issues of environmental protection and ecological integrity cannot be divorced from important and conflict-ridden questions of human rights (e.g., Johnston 1995).

This connection between environment and human rights can cut both ways, of course. In particular, indigenous peoples are increasingly bearing the brunt of the often naive agendas of environmentalists living in urban centers far away. For example, as Beach (chapter 6) details, Saami reindeer herders in Swedish Lapland are coaxed and regulated into "rationalizing" their production system, and then attacked by environmentalists for abandoning their traditions and sacrificing ecological integrity for monetary gain. Copper Inuit hunters in Canada's Central Arctic are treated like delinquent children by a wildlife biologist who knows far less about caribou populations than they (Collings, chapter 2). These two instances mirror other well-known cases in which environmentalist and/or animal-rights campaigns have seriously damaged the subsistence production of northern Natives, such as the anti-seal-hunting initiative spearheaded by Greenpeace and other NGOs (detailed in Wenzel 1985, 1986), and the attack on subsistence whaling (see Freeman 1994; Freeman and Kreuter 1994).

At the other end of the spectrum are instances of cooperation between Natives and non-Native environmentalists. This volume contains at least two clear examples of such alliances. One is the strong support provided by environmentalist and Native-rights activists to the Lubicon Cree in their battles with timber interests (Espiritu, chapter 3). The other is the indication Johnson (chapter 1) gives that Greenpeace (among other conservation groups) supported an exception for Native subsistence use in the reauthorization of the U.S. Marine Mammal Protection Act; this is striking because of the hostility which Greenpeace previously expressed toward any Native subsistence exemption in the seal-hunting controversy of the 1980s.

Clearly, there are ethical as well as practical and empirical issues that must be faced here. A view commonly expressed by those who defend Arctic industrialization or resource extraction is that the interests of a few (i.e., indigenous peoples) cannot be allowed to limit the greater good of resource utilization for the many (i.e., the workers, consumers, and national economies of the colonizers). Thus, as Johnson (chapter 1) recounts, former Alaska Governor Wally Hickel has argued that since it is devoid of people, Alaska is rightfully viewed as a storehouse of resources for entire U.S.A. How could anyone justify denying increased electrical power or gasoline to Montréal, Anchorage, or Vladivostok for the fear of harming caribou herds somewhere in the "trackless" north, or in an attempt to protect a "primitive lifestyle" that is doomed anyway? Such arguments closely resemble the justifications propounded by European colonizers invading the Americas five centuries ago, who justified their appropriation of indigenous lands and resources by claiming the local inhabitants weren't using them them to the fullest extent possible (Berger 1993). Of course, this crude utilitarianism is not so readily accepted if one applies it to housing for homeless people in empty hotels or "underpopulated" homes of the wealthy—or even in public parks!

The utilitarian rhetoric can of course be inverted: Is the energy leaking from a million faulty carburetors or drafty windows or all-night supermarkets worth the destruction of entire ancient cultures and degradation of entire ecosystems? And how can the urban civilizations colonizing the Arctic, already awash in material goods, justify destroying the homelands and livelihood of northern hunters and herders who consume so little themselves? Ultimately, these are moral issues, not logical or empirical ones. In the present volume, it is quite clear where the authors stand on these issues—no "value-free" social science here. Of course, we must guard against romanticizing indigenous peoples, or portraying them solely as victims: like people everywhere, they sometimes choose economic gain over environmental preservation, or factory jobs over traditional subsistence. But as accounts in this volume show, their choices are usually limited, and they regularly get the worst of both worlds.

Resistance, Reform, and Other Glimmers of Hope

Clearly, the Arctic's position as a colonial frontier of resource extraction and cultural conflict has produced a serious set of social and ecological problems. But the narratives in this volume are not uniformly or unrelentingly grim. We find, for example, that Inupiat hunters in Northern Alaska are linking up with Chukchi reindeer herders across the Bering Strait to help forge bilateral Russian-U.S. agreements on conservation and subsistence harvests of polar bears and other marine mammals (Johnson, chapter 1). The collapse of the Soviet Union has opened a huge new window for the West to learn about environmental pollution, industrial development, and struggles to preserve or legislate Native rights in Siberia and beyond (chapters by Fondahl, Espiritu, and ZumBrunnen). Indeed, one of the things that makes this book so timely is the way it draws on the recent upsurge in information about the vast swath of circumpolar environment and peoples that extend across Russia's north.

The weak political position of indigenous peoples in the contested present and future of the Arctic flows from several factors: low numbers, little capital, and scarcity of information. While the first two disadvantages are unlikely to change much in the near term, northern peoples are increasingly finding ways to gather strategic information and put it to political use. Some indigenous rights activists argue that access to the Internet and manipulation of international media and political bodies (e.g., the United Nations) is the best hope for halting the colonial juggernaut in the forests of Borneo and along the coasts of the Bering Sea. Whether true or not, it is remarkable to learn from Charles Johnson (chapter 1), himself a Bering Sea Native, how Arctic peoples are taking an active role in gathering and analyzing information on transboundary pollution, global warming as monitored via Greenland icecap cores, and the epidemiological and subsistence impacts of persistent organic pollutants.

But obviously information is not enough. This information must be tied to coherent plans of political action, and it must be employed strategically to inform people within and beyond the Arctic concerning the social and ecological consequences of their actions. Again, Johnson's chapter indicates how strategic information is coupled with carving out a role for indigenous peoples themselves in policy formation and enforcement. There is the danger that indigenous participants on bilateral regulatory commissions or monitoring agencies will become bureaucrats with their own interests, with agendas increasingly remote from the daily lives and concerns of their relatives and neighbors in northern villages, traplines, and fishing camps. Given the stakes and the disadvantages under which indigenous populations currently labor, this seems a risk that must be taken.

While the chapters of this volume are filled with particulars of various places and peoples, one ironic conclusion we can draw from them is how parallel are the fates of both ecosystems and indigenous peoples throughout the circumpolar north, from the welfare state of Sweden to the "communist empire" of the former Soviet Union to the capitalist nations of Canada and the U.S. Nowhere is this common fate more apparent than in the devastating comparison Aileen Espiritu draws between the Lubicon Cree of Alberta's oil fields and the Native peoples on the industrial frontiers of Siberia. In the final analysis, it seems to matter little whether the colonial powers intent on developing the north are state-socialist, social-democratic, or avowedly capitalist, or whether their governmental structures are based on parliamentary democracy or the dictatorship of the proletariat. In all cases, the imperatives of development and resource extraction proceed at the expense of local inhabitants. Perhaps this is because of one thing these colonial contexts all have in common: the indigenous inhabitants are ethnic minorities, their lands overrun by invaders who generally judge that the Natives have no effective systems of resource utilization nor any valid ecological knowledge, and hence feel fully justified in their imposition of external models of resource management.

These parallels are very instructive concerning the historical processes that shape the political and environmental problems afflicting the circumpolar north. We can hope that they may also suggest common routes towards solving or ameliorating these problems, while there are still viable indigenous cultures and functioning Arctic ecosystems left.

References

Berger, Thomas R. 1992. *A long and terrible shadow: White values, Native rights in the Americas, 1492-1992*. Seattle: University of Washington Press.

Burch, Ernest S. 1994. Rationality and resource use among hunters. In *Circumpolar religion and ecology: An anthropology of the north.* Takashi Iromoto and Takako Yamada, eds. 163-185. Tokyo: University of Tokyo Press.

Chance, Norman A. and Elena N. Andreeva. 1995. Sustainability, equity, and natural resource development in northwest Siberia and Arctic Alaska. *Human Ecology* 23:217-240.

Clay, Jason. 1993. Looking back to go forward: Predicting and preventing human rights violations. In *The state of the peoples*. Marc Miller, ed. Boston: Beacon.

Crosby, Alfred W. 1986. *Ecological imperialism: The biological expansion of Europe 900-1900*. Cambridge: Cambridge University Press.

DINA. 1976. *The James Bay and Northern Quebec Agreement*. Ottawa: Department of Indian and Northern Affairs, QS- 5051-000-EEAI.

Freeman, Milton M.R. 1994. Science and trans-science in the whaling debate. In *Elephants and whales: Resources for whom?* M.M.R. Freeman and U.P. Kreuter,eds. 293-300. Basel: Gordon and Breach.

Freeman, Milton M.R. and U.P. Kreuter, eds. 1994. *Elephants and whales: Resources for whom?* Basel: Gordon and Breach.

Johnston, Barbara Rose. 1995. Human rights and the environment. *Human Ecology* 23:111-123.

Langdon, Steven J. 1989. Prospects for co-management of marine animals in Alaska. In *Cooperative management of local fisheries.* Evelyn Pinkerton, ed. 154-169. Vancouver: University of British Columbia Press.

Norgaard, Richard. 1988. The rise of the global exchange economy and the loss of biological diversity. In *Biodiversity.* E.O. Wilson, ed. Washington, D.C.: National Academy Press.

Schneider, Howard. 1996. Nations begin to tackle Arctic pollution. *Seattle Times,* September 22, page A18.

Smith, Eric Alden. 1991. *Inujjuamiut foraging strategies: Evolutionary ecology of an Arctic hunting economy.* Hawthorne, NY: Aldine de Gruyter.

Thouez, J.P., A. Rannou, and P. Foggin. 1989. The other face of development: Native populations, health status, and indicators of malnutrition—the case of the Cree and Inuit of northern Quebec. *Social Science and Medicine* 29:965-974.

Usher, Peter J. 1987. Indigenous management systems and the conservation of wildlife in the Canadian North. *Alternatives* 14:3-7.

Wenzel, George W. 1985. Marooned in a blizzard of contradictions: Inuit and the anti-sealing movement. *Etudes/Inuit/Studies* 9(1): 77-92.

———. 1986. Canadian Inuit in a mixed economy: Thoughts on seals, snowmobiles, and animal rights. *Native Studies Review* 2(1): 69-82.

Young, Oran R. and Gail Osherenko. 1993. *Polar politics: Creating international environmental regimes.* Ithaca, NY: Cornell University Press.

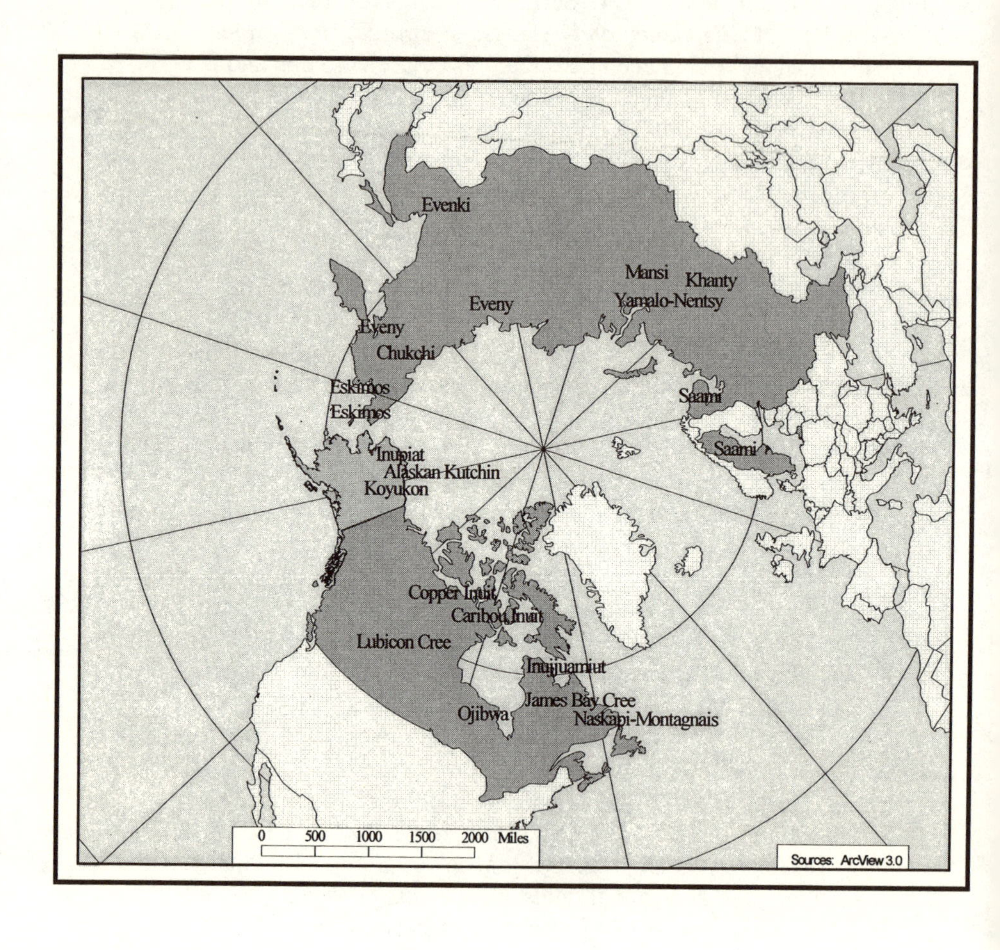

MAP 1 Indigenous Peoples of the Arctic

Contested Arctic

Indigenous Peoples, Industrial States,
and the Circumpolar Environment

1

The Role of Indigenous Peoples in Forming Environmental Policies

Charles Johnson

I would like to begin with a poem by Ed Edmo, an Indian poet, called "the Mountain":

> *I like that mountain over there*
> *I don't want to walk on old man mountain*
> *I don't want to dig holes in his face*
> *I don't want the smoke to get in the way of the sunshine, for*
> *I believe that mountain likes sunshine*
> *Just like me*
> *But most of all*
> *I don't want that mountain to get mad at me*

The poem reflects our feelings that we, as Native people, are part of the Arctic ecosystem. We are not observers, not managers; our role is to participate as a part of the ecosystem. This is the basis for our outlook on the environment and the primary consideration when we look at forming an Arctic policy.

Our first concern is global warming. This is something over which we have little or no control or impact, but something that will have the most dramatic impact on us and the Arctic environment. When I was a member of the U.S. Arctic Research Commission, we received reports and data from the Greenland Ice Sheet Project (GISP) on historical climatic changes over the last 100,000 to 150,000 years. By drilling from the top of the ice sheet to bedrock and analyzing the ice core, the scientists can very accurately discern and date temperature and climatic conditions, as well as catastrophic events such as volcanic eruptions. The preliminary data show that there is a warming and cooling cycle of approximately 1,500 years, perhaps caused by astronomical conditions. Generally there is a very slow rise in temperature followed by quick rise before the temperature drops again. The weather data show that there has been a dramatic rise of six or seven degrees Celsius in the Arctic over the last fifty years.

One has to question if this rise in temperature is the result of the cycle shown by GISP, or if it has been affected by the introduction of greenhouse gasses into the atmosphere by man. It very well could be that human activity has accelerated the historic cycle. These studies and others have shown that the temperature fluctuations in the polar regions are not paralleled by similar rises and falls of temperature in the tropics and temperate zones, where the temperature changes are very small, between 1° and 2° Celsius.

The most obvious effect of rising temperatures will be the rise of ocean levels as water melts from polar ice caps and glaciers. For the U.S. and other parts of the temperate zone, the rising oceans will flood low-lying areas like Florida, Maryland, and maybe even parts of Seattle. Thus, the concern over global warming at the policy levels of government is not so much a concern for the Arctic but for the effects like flooding that global warming will have on the heavily populated areas of the U.S. The government gives us lip service, but our population is not big enough to have much political impact.

For those of us living in the Arctic, particularly indigenous peoples, global warming will have a more dramatic effect. It will change the way we live in ways that we cannot now envision. Global warming will change the plants and the animals and their migration patterns. It will change the subsistence cultures of the Arctic.

Our second concern, again one over which we have no control, is transboundary pollution. We know that persistent organic pollutants (POPs) are among the pollutants tracked by the Arctic Monitoring and Assessment Program (AMAP). AMAP is one of the working groups of the Arctic Environmental Protection Strategy (AEPS), an initiative of the eight Arctic nations: Canada, Denmark/Greenland, Finland, Iceland, Norway, Russia, Sweden, and the United States.

We are starting to pick up POPs and other pollutants such as heavy metals in the meat, organs and fat of the animals that we use for subsistence in the Arctic. I have been taking samples for pollutant analysis from seals, walrus and beluga whales in the Norton Sound of the Bering Sea where I live. We take blubber, liver and kidney samples. We are looking for POPs and for heavy metals. We are picking up heavy metals such as cadmium in walrus livers. AMAP is picking up large amounts of POPs in other areas of the Arctic.

In the case of heavy metals, our anxiety is raised by questions of what is natural and what is caused by man's industrial activity. We know for example that cadmium is naturally occurring. But when we see differences in the samples, particularly from the western Bering Sea close to Russia, we worry that the cadmium is coming down the Russian rivers. Walrus do not know where the boundaries are; they travel visa-free back and forth. Geographic differences in the samples creates a worry.

We know that the sources of transboundary pollution, particularly POPs, are in the temperate areas of the world. Even though many of these pollutants, particularly DDT and other related pesticides, are banned in the U.S., the U.S. is now manufacturing more DDT than when it was first banned. It is shipping DDT to Third World countries, where it enters the atmosphere precipitates out in the Arctic. So the U.S. itself is the source of these dangerous organic pollutants that are causing us so much concern.

When POPs are discussed by the people of the Arctic the first concern is always cancer, since it is widely known that POPs cause cancer. But we are now finding out that cancer is only the final problem, and it may be the least of the problems. After participating in several workshops on the ecological effects of pollutants, we are finding out that while cancer may show up in the persons exposed to these pollutants, the real problems are showing up in their offspring. Their children are having reproductive problems, as well as psychological problems such as depression and cognitive problems. And these problems take years to become apparent.

The Alaska Native Medical Center and the State of Alaska Health Department have recorded a dramatic shift in the causes of death among Alaska Natives. Fifty years ago, as is true now, accidents were the number one cause of death, including drowning, freezing, fishing accidents and so forth. But fifty years ago the second-most common cause of death was infectious diseases, such as tuberculosis, measles, etc. Today, it is heart problems. Cancer is third. It has become apparent that the shift in the causes of mortality is largely a result of a change in diet and lifestyle.

Cancer scares people. When we see that it is one of the leading causes of death and that the occurrence is rising we take notice. It seems insidious because it seems to appear out of nowhere. Perhaps that is why we are more concerned with the cancer that we attribute to pollutants than with the other effects that these pollutants may have. While the reports seem to downplay the role that pollutants have in mortality, we must be concerned with the incremental shifts in the environment that these pollutants cause, and the eventual cumulative harm that is being done.

We know from Peter Collings's chapter that there is a shift in the diet of the Native people in the Canadian Arctic. The same change is evident in Alaska. The explosion of television and communication has created a desire for material things and thus for cash. This means jobs. The need for jobs and the educational requirements to get these jobs mean a move away from a subsistence lifestyle. Young people now must spend time in school and don't learn as much about the healthy traditional foods like greens, herbs and roots that have met our nutritional requirements in the past. Consequently, there is a shift to store-bought, processed foods with their resulting health problems.

These health problems are showing up in my generation. Even though I prefer seal, walrus and other traditional foods, my diet includes bacon and other easily bought processed foods. Consequently, even though I consider myself active and healthy, I have had heart problems. I just got out of Providence Medical Center in Anchorage, where I had an angioplasty to clear a clogged artery.

Because of the demands of this modern society that our children are moving into, the trend toward unhealthy diets will continue. We aren't able to teach our children about all the traditional foods, particularly greens and roots, that kept us healthy in the past. While my generation was healthy because of these foods, our kids haven't learned of their practical benefits. It will take a massive and protracted educational effort on our part to learn which of the store-bought foods are good for us and which ones to avoid.

Another area of concern we have is industrial development in the Arctic. We are concerned that oil development, mining and the resulting infrastructure will have an effect on the environment and on the species that we rely on. What will the effect be on the migration patterns of caribou, migratory birds and other species? Will there be effects on salmon and other fish? We observe shifts in patterns but, to date, government officials are able to dismiss these observations as unscientific, anecdotal, or biased. This change in our resource patterns contributes to the shift in our diets and habits.

A major example of development that is having a dramatic effect is the impact on subsistence from commercial fishing. Right now in my area of Alaska we have not had a subsistence chum salmon season for six straight years. The reason given us is that we must have an escapement up the rivers for the salmon to spawn. But the fish are not being caught by our meager subsistence nets. They are being caught in huge nets by commercial fisherman at False Pass, and the State of Alaska Fisheries Board keeps increasing their allowable catch of chum salmon. These commercial fishermen, who are mostly from Seattle and other non-Alaskan cities, are making huge amounts of money, averaging $250,000 to $400,000 per share over the same six years that we are being denied subsistence for the sake of saving the same salmon run for the benefit of these out-of-state commercial fishermen.

Our people are seeing a dramatic decline of species in the Bering Sea. Right now the Bering Sea is being strip-mined for fish by huge trawlers that are taking everything off the bottom of the sea. They are only allowed to keep the target species. If they are fishing for pollock for example, they must dump overboard the salmon, halibut, crab and other species that are brought up in the nets. The reports of what is dumped is kept by the fishermen themselves. By their own reports, they have dumped an average of between 700 million and a billion pounds of these highly valuable fish every year. And this is only what they have reported—we feel the actual amount is much higher. It is no wonder that we are seeing a decline of some species in the Bering Sea.

We have reported a decline as high as eighty-five percent in Steller Sea Lions and sea birds in the Bering Sea. This decline hasn't been apparent in walrus for at least two reasons: the walrus spends much of its time in the Chukchi Sea where there is no bottom-fishing; and the food of the walrus, clams and other bottom food, is not commercially exploited in the Bering and Chukchi Seas.

Another environmental concern is one I will call cultural, for lack of a better term. I've talked about our younger people not being able to go out on the land, not learning many of the skills that are needed to live in the Arctic by hunting and gathering, because the law says our kids have to have a formal education. For example, when I was a kid, I had to go to school. My parents didn't speak English when they were sent to school, so they made me speak English so I wouldn't be punished like they were. I was left-handed, but the teachers made me write right-handed because that's what Americans are supposed to be I guess. But again I'm reminded of a poem by Ed Edmo. He is a Shoshone Bannock from Idaho who wrote this poem when he was in school in the 1950's.

I sit in your crowed classroom
And learn to read about Dick, Jane and Spot
Yet, I can remember how to catch a deer
I can remember how to do beadwork
I can remember the stories told by the old
But Spot keeps turning up
And my grades are bad

It is this cultural environment I am referring to when I say that our environmental concerns are not only physical. The cultural environment shapes what we are turning into. It has gained primacy over the physical environment which shaped us in the past. Perhaps the biggest reason is immediate and ubiquitous communication by television. Television is the biggest cultural pollutant that we now have. It has overcome organized religion in that respect. Every village in Alaska now has satellite or cable television. Everyone spends their time watching television. I have a cousin in my home village that doesn't send me ice-caught fish anymore, because he spends his time watching TV rather that fishing through the ice.

Religion still has a powerful hold in some areas of Alaska. For example, in the Northwest Alaska Native Association region, if you are not a member of the Society of Friend's Church, it is very difficult to be elected to any board, council or public office. In effect, we have a church controlling a region of Alaska. I used to have a map that was drawn up by the National Council of Churches in 1913 to divide up Alaska between the various denominations. You can still tell which part of Alaska a Native comes from, by their religion.

Political and economic influences are playing a big role in environmental policy for Alaska and the rest of the Arctic. Alaska Natives are approximately seventeen percent of the population of Alaska. Our political influence is growing smaller as more and more non-Natives are moving to Alaska. They in turn want to shape Alaska into a copy of their last residence, whether Seattle, Dallas or Oklahoma, much as the English pilgrims that made New York, New England, etc. As a result, our legislature is more anti-Native and anti-subsistence. Natives and subsistence, they feel, deny them their "God-given right" to use and control the environment. Fortunately, in federal legislation, subsistence has priority over other uses of wildlife in rural areas of Alaska. Because the State of Alaska legislature will not recognize a subsistence priority, the federal government has regained control over fish and game management on federal lands, which is sixty percent of Alaska.

State legislators will not let the people of Alaska decide the subsistence priority issue. Instead they have attempted punitive measures such as calling for the audit of Native Corporations, and attempting to deny state funds to communities that recognize Tribal governments. One of our past governors, Wally Hickel, who to some is just to the right of Attila the Hun, has stated that Alaska is just a storehouse of resources for the United States, because it is devoid of people. Where there are resources only a few Natives live, and the resources need to be developed for the benefit of the United States. At the same time, he appointed a thirty-three year-old surfer from Florida to represent the State of Alaska to other circumpolar governments on the Northern Forum, an international organization of leaders from northern and Arctic regions. This is the type of political situation we are facing.

There is a popular misconception in Alaska and the rest of the U.S. that we Natives received a large hunk of Alaska. When the government of Alaska was transferred from Russia to the U.S. in 1867 a clause stated that the land belong to the Native people, and that we would be compensated if we lost the land. Congress, in the Alaska Native Claims Settlement Act (ANSCA), stated that we lost 320 million acres, including Prudhoe Bay and other oil-rich areas, and that we were being paid $3 an acre for what we lost, but we could keep 44 million acres of land that was already ours. The general population is still under the impression that we were given 44 million acres, while in reality we lost 320 million acres and all its resources.

As we in Alaska are becoming more and more politically astute, we are learning that the problems we face are not unique to Alaska. The Natives of the circumpolar north are facing the same problems. To combine our strength and share our support, the Inuit Circumpolar Conference (ICC) was formed. In 1977, Eben Hopsen, the mayor of the North Slope Borough, called a planning meeting of the Inuit of Alaska, Canada, and Greenland to develop the ICC. In 1980, the ICC was actually formed in Nuuk, Greenland. The ICC has been accepted as a non-governmental organization by the United Nations, which allows us to take our concern to the international community. In 1992, the Inuit of Russia became members.

When we were concerned only about our provincial problems, which were and still are major, our whipping boy was always the Bureau of Indian Affairs (BIA) in the U.S. government. We are now taking over the functions of the BIA in providing services to our people. Since we have now largely phased out the BIA and have moved into the international arena, our attentions are drawn to the shortcomings of another U.S. agency—the State Department. They now are one of our "whipping boys." I'll explain.

I noticed in the statement presented by Molly Fayen* that the U.S. State Department's policy on the Arctic supported the involvement of indigenous "people." Notice that "people" is singular. Apparently the State Department feels that indigenous people (singular) are all the same. They will not use the term "peoples" (plural), for to do so would recognize that we are not all the same. For this and other idiotic actions, the State Department has replaced the BIA as my favorite whipping boy. This is not meant as a personal attack on Molly, since she is only representing the State Department.

Our engagement with the U.S. State Department began with ICC's fighting to become involved with the Arctic Environmental Protection Strategy, a process begun by Finland in 1989. Finland recognized that the Arctic environment was at risk and called the eight circumpolar nations together to discuss a strategy. The ICC and other indigenous organizations such as the Nordic Saami Council, and the Northern Indigenous Peoples of

---

*(Ms. Fayen, a Polar Resource Officer at the U.S. State Department, presented a paper entitled "Intergovernmental policy coordination: The Arctic Environment Protection Strategy." Despite our invitation, she declined to submit her paper for inclusion in this volume. —eds.)

the Soviet Union, pushed for a seat at the table, stating that the peoples who relied on the Arctic environment must be involved. The U.S. State Department was very much against our participation, stating that we are just an interest group like the environmental organizations. Fortunately, the other nations agreed that our involvement was needed.

When the AEPS process started, I was representing ICC and I was the only Alaskan there. I was outside the U.S. delegation, which was primarily from Washington, D.C. Those developing Arctic policy for the U.S. were not very familiar with Alaska. In fact, one person working on the White House Arctic Policy spelled my home town of Nome as Gnome.

The major environmental organizations have been very important to the indigenous groups. During congressional reauthorization of the Marine Mammal Protection Act, Greenpeace and other conservation groups supported a Native exception to the prohibition of taking of marine mammals so that we could continue our subsistence way of life.

The MMPA called for a bilateral treaty with Russia for the conservation of the shared population of polar bears in the Bering and Chukchi Seas. We were invited by the U.S. Fish and Wildlife Service to observe the process. We responded that because the U.S. had agreed to the AEPS process we wanted to not only participate, but we wanted equal participation. We created the Alaska Nanuuq Commission to form an organization to represent the polar bear villages in Alaska from King Island in the Bering Straits Region to Kaktovik on the Canadian border. We also helped the Natives of Chukotka in Russia to form their own organization, UMKY, the Polar Bear Commission, Chukotka. (*Umky* means *nanuuq* or polar bear in the Chukchi language.)

The process started when Russia informed the U.S. that they were taking the polar bear off the "Red Book" list of endangered species, and reclassifying the polar bear from endangered to recovered, so that they could be legally hunted. Presently, we know that polar bear are being taken in Russia, mostly illegally and not all by Native hunters.

There are two agreements between the Russian and U.S. governments. The first of these:

- creates an umbrella agreement between the governments that recognizes exclusive use of polar bears by the Native peoples (we are fighting for the "s" in peoples with the State Department) for subsistence purposes;
- creates a Joint Commission of four members, one from each of the governments and one from each of the Native Commissions which will operate on a consensus basis; and
- recognizes an agreement between the Native Commissions which will be the implementing agreement.

The second agreement is between the Alaska Nanuuq Commission and UMKY, which will allocate a quota, if necessary, and enforce the provisions of the agreement.

It is important to note that without the work done by ICC in the AEPS process, we might not have had an equal voice in this process. The movement to become involved in the international arena is critical to preserving our choices and political influence.

The AEPS is now evolving into an Arctic Council, which is comprised of the same eight circumpolar nations and three indigenous umbrella groups: the ICC, the Nordic Saami Council, and the Northern Indigenous Peoples of Russia. Initially, the Arctic Council will address environmental issues, but will also later address other circumpolar Arctic issues.

It is important to also note that the Arctic Council was first proposed by the ICC president, Dr. Mary Simon, who later became Canada's Arctic Ambassador. Formation of the Arctic Council was agreed to with great reluctance by the U.S. government, which felt that indigenous groups could not be allowed to sit as equals with governments on policy-level issues.

The bilateral agreement we are working on with Russia is illustrative of the type of government attitudes we have worked hard to overcome in Alaska and the rest of the free world. Russia is now just opening up and is decades behind the rest of the Arctic in its relations with indigenous peoples.

I spoke at the first Congress of Indigenous Peoples of Chukotka in Anadyr, Chukotka in April, 1994. The speaker before me was Governor Nazarov of Chukotka, who is Kazak by descent. He wanted to know why Native people needed a special "Congress," stating "what in the hell do you Native people think you want? What is your problem? What makes you think you are special? There are Russians here, Kazaks, Ukrainians who all came here for your benefit, to open the mines and develop this area for your benefit." (Does this remind you of an Alaskan Governor who views Alaska as a storehouse?) The governor further stated that they have Departments of the Problems of Minorities, of the Problems of Health, etc. and that these departments should adequately address the concerns of Native peoples.

The sad fact is that governments, including that in Alaska, regard Natives as a problem. Government attitudes toward indigenous peoples are a "problem" that is common across the Arctic. As in Alaska, the federal government in Moscow recognizes indigenous peoples as unique and has to date agreed to the concurrent agreements on polar bear. Both the state governments of Alaska and Chukotka are against having a "Native-to-Native Agreement" because it denies non-Natives their "rights" to be involved in our affairs. Also, like in Alaska, the regional government would like to control the renewable resources like fish and game. The bilateral treaties like the polar bear treaty are essential to protect the rights of indigenous peoples.

We would like to develop similar agreements with Russia on other shared marine mammals like walrus, seals and whales. These species pose other problems that are more complicated than that of the polar bear. Walrus, for example, are hunted in Alaska by villages with no limits as long as there is no waste. In Chukotka, the fisheries inspectors issue quotas to villages for a specific number of walrus annually. The quotas are really meant for the fox farms who have the hunting equipment, not for the Native people. When I was in a meeting with Native people and fisheries inspectors and the Natives were interested in an agreement since they were having trouble getting meat, the fisheries inspector stated that they could just go buy walrus meat that the foxes wouldn't eat. For an individual to hunt a walrus, he first has to get a license, then check out a gun and ammunition and then get a date-specific permit from the Border Guard to leave on a hunt at sea. The bureaucrats are having a tough time letting go of the people.

For us in Alaska, what happens all over the circumpolar north affects us. If we are going to protect our environment and the food it provides us we have to make sure it is protected all over the Arctic. What goes down in Russia or Canada eventually will have an impact on us.

We were born to an environment we thought was isolated from the problems of the rest of the world. But we have found out we cannot ignore what happens outside of our borders. Our role is to be stewards of our environment, since we are only keepers for the next generations.

2

The Cultural Context of Wildlife Management in the Canadian North

Peter Collings

In 1992-1993, I had the opportunity to work as a research assistant for the late Richard Condon in Holman, a settlement of Copper Inuit located on Victoria Island in the Northwest Territories of Canada. Ostensibly, I was there to collect data on subsistence hunting and economic adaptation with a sample of young adult household heads. While conducting this research, I observed the emergence and apparent resolution of a caribou crisis in the community. Throughout the course of my stay in Holman, hunters had noted with concern that caribou hunting was less and less productive, and by early 1993 many had ceased hunting caribou altogether. At about the same time the hunters were stopping their caribou hunting, a non-Native wildlife manager arrived to study the problem. A local Inuk was hired to interview community elders about the problem, and the manager made several aerial surveys of the region to estimate caribou population numbers. Several months later, the manager presented the results of the study to the Holman Hunters' and Trappers' Association and argued that local hunting was largely responsible for the decline of the caribou population. Based on the manager's recommendations, a hunting ban was instituted by the government, with provisions for the ban to be reconsidered after three years (in spring, 1996).

The research in Holman discussed in this paper was supported by a grant from the National Science Foundation (DPP grant # 9110708). I am also grateful to Patricia Draper, George Wenzel, and especially Eric Smith for helpful insights and comments on earlier versions of this paper.

The sequence of events and its outcome was anything but a heartening picture of wildlife management. It was clear from the outset that the manager strongly disliked the local Inuit and sought as little contact with them as possible. Consequently, Inuit had little if any input in either conducting the research or interpreting the results. The evidence supporting hunting as a primary cause of population decline was less than convincing to those familiar with standard scientific procedure. The manager virtually ignored local knowledge of caribou and their movements, which suggested that caribou had simply moved away in search of better forage. Some hunters later felt vindicated by the reappearance of caribou during the winter of 1994-1995, when caribou returned to the region and began venturing close to the settlement (Harold Wright, personal communication, March 1996), and the ban has been conditionally lifted (Louie Nigiyok, personal communication, May 1996).

Equally impressive was the reciprocal dislike that a number of elder Inuit had toward the manager. Not only did these elders think that the manager's knowledge of caribou was questionable, some challenged the validity of the methods used in conducting the study. A few noted from personal experience that aerial surveys of caribou are not always accurate because the animals blend in with their surroundings, for example. Others were concerned that wildlife managers were in general out to get Inuit for their own sinister purposes: "The government is always taking money, stealing money and using it for themselves," as one elder put it.

This paper is an examination of some of the underlying assumptions of wildlife management in the Canadian North—namely, that people cannot manage their resources and are subject to the "tragedy of the commons." An examination of this paradigm and some of the criticisms of it will be followed by several examples of cases where local, indigenous management systems are known to have operated effectively over time. Conditions under which local management regimes fail will also be discussed. The final section of this paper will address the political context of wildlife management, focusing on the following issues: 1) the merits of both science and traditional knowledge in understanding local ecology; 2) the reluctance of wildlife managers to consider local, traditional knowledge when making management decisions; and 3) the effectiveness of incorporating knowledge derived from the perspective of wildlife management and local, traditional knowledge in pursuit of what is widely known as co-management (Berkes 1981a).

The Tragedy of the Commons

Wildlife managers generally believe that without external coercion, people are incapable of collective action to manage environmental resources for the common good. This is a world view rooted in the philosophy of Hobbes (see Berkes and Feeny, 1990, for a lengthy discussion). The underlying assumption that follows from the Hobbesian view seems to be that so-called primitive peoples did not manage their own resources because primitive technology prevented efficient hunting, which in turn forced peoples to follow a nomadic lifestyle. Taken together, these two factors prevented people from observing the consequences of their actions and making appropriate adjustments (Theberge 1981). Indeed, human foraging prior to contact with Europeans is thought to be best characterized as a complex predator-prey interaction, where human predators are prevented from exacting too great a toll on their prey by their limited technology. If this limit is not sufficient, then overharvesting is counteracted by starvation (Freeman 1989; MacPherson 1981). In cases where resources are susceptible to overexploitation, foragers are thought to have shown little restraint and hunted their prey to local extinction, often through wasteful practices. Oft-cited examples include caribou hunting on the Barren Grounds (Kelsall 1968) and local muskoxen extinctions (noted by Burch 1986 and Stefánsson 1913, among others). This Hobbesian view is elaborated in a popular paradigm known as the tragedy of the commons.

The tragedy of the commons was first outlined by Gordon (1954) in reference to fisheries use, but it did not become ubiquitous among wildlife professionals until Garrett Hardin expanded its scope in his now famous 1968 paper. Hardin advanced his argument with reference to a hypothetical group of herders using a common pasture on which they graze their animals. On this pasture, Hardin assumed that each herder could sustainably graze x animals. Increasing any herd above x would eventually degrade the pasture until, at some point in the future, it could no longer support any grazing. Each herder, however, also realizes that it pays to add extra animals. While adding an animal is detrimental in the long run, the individual herder who does so reaps the benefits derived from the extra animal (increased production), while the costs of adding it (ultimate degradation of the pasture) are shared equally among all users. Since each herder comes to the same realization, it not only pays to add additional animals to the pasture but also becomes necessary since refraining in the name of conservation means that the conserver shares only in the costs of overgrazing but gains no benefit. The tragedy is that each individual, by behaving in a rational manner, is locked into a ruinous path of environmental degradation and, ultimately, starvation. The argument has been applied extensively to explain the causes of famine in both East and West Africa (Fratkin 1991:12-13), overfishing (McCay 1987), and wildlife overexploitation (Freeman 1989).

The logic of the tragedy of the commons is quite compelling, but is it true? Examinations of Hardin's formulation have questioned the paradigm on numerous grounds, and critics have been left wondering if the tragedy of the commons exists at all. Feeny and colleagues (1990) have suggested that Hardin's argument confused the issue of the tragedy of the commons because he misunderstood the nature of property rights. Feeny et al. define four different property rights regimes: 1) open access, where specified rights of access are minimally defined or nonexistent and there are no mechanisms for excluding others; 2) private property, where well-defined rights of access and mechanisms for excluding others are vested in the hands of an individual entity (either a single person or a corporation, for example); 3) communal, or common, property, where the use of the resource and mechanisms for exclusion are held by a group of interdependent users; and 4) state property, where resource rights are held by the government, which administers and limits access to its use. Hardin argued that the best ways to resolve the dilemma were through privatization or through mutually agreed-upon coercion enforced by the state. One of the primary problems of Hardin's argument is that he recognized only three property rights regimes and did not distinguish between communal, or common property, systems and open access systems of land tenure. Under common property arrangements, the resource is shared by a group of interdependent users, who exist within a cultural context that dictates who has access to the resource, when they may use it, and how they may use it (Swaney 1990). In his formulation, Hardin assumes that each herder has no social contact with any other herder. Indeed, his only mention of a social context is as follows:

> Such an arrangement [herding on a commons] may work reasonably satisfactorily for centuries because tribal wars, poaching, and disease keep the numbers of both man and beast well below the carrying capacity of the land. Finally, however, comes the day of reckoning, that is, the day when the long desired goal of social stability becomes a reality. At this point, the inherent logic of the commons remorselessly generates tragedy (Hardin 1968:1244).

Hardin's statements about carrying capacity, tribal wars, and disease reinforce assumptions about so-called primitive humans. The important point, however, is that Hardin ignores cultural and social context as important influences on patterns of resource use. Hardin was therefore describing not a tragedy of the commons, but a tragedy of open access. A second criticism of Hardin also focuses on the ignorance of social context. Berkes (1985) has noted that there are three conditions necessary for the tragedy of the commons to appear: 1) users must be selfish and able to pursue their own interests at the expense of the group; 2) there must be a limited environment such that resource use exceeds resource regeneration; and 3) the resource must be collectively owned so that each member has access to it. As Berkes points out, once the conditions are met, the tragedy becomes inevitable, and we are left with a tautological argument. The important question is how social and cultural contexts can affect the emergence or suppression of conditions that lead to a tragedy of the commons. Or, stated more simply, "Do indigenous peoples effectively manage their resources?"

Indigenous Property Management Systems

The "Conservation Ethic"

A popular belief about Native peoples is that they are "natural conservationists" or the "original ecologists," people who see themselves as stewards of the land and living in harmony with nature. Native religious ideology and spiritual beliefs likewise serve to reinforce the conservation ethic by requiring appropriate behavior and attitudes toward animals and providing supernatural sanctions for violating these requirements. Taken together, cultural values maintain the balance between humans and nature. Among the Copper Inuit with whom I work, for example, elders often speak of the importance of not talking about animals before going hunting, as animals can hear people and can become offended if they are not treated properly. Bears are thought to be especially sensitive to human conversation, and talking about bears is thought to either upset them and cause them to move away or to appear to exact revenge on the transgressor, as the following excerpt from Condon's field notes indicates:

> Peter [Collings], David [a pseudonym], and I [R. Condon] went out fishing and musk ox hunting yesterday. We left around noon and first went to Pocket Knife Lake to fish. The ice was pretty thick, so were able to drive our machines over the lakes. Louie got one good-sized lake trout. We then drove around for awhile, fishing at other lakes and looking for musk ox. We met three ski-doos at one of the lakes: Alec and Mary Kittekudlak, William Kittekudlak, and Sarah Kanayok. They were all driving around and fishing like us. We fished with them for awhile and had tea. At one point, I was having tea and talking with Mary, Sarah, and William. They called Peter over for tea. William said jokingly to Peter that there was a polar bear behind him. Sarah chimed in and said that it was about this big, holding her hands up to show a really small bear. Sarah then shuddered and said that we shouldn't talk about animals like that because they might hear you and come to you when you don't expect them. She then asked me if I had heard about what happened to Jimmy last

> summer when he was attacked by wolves. I said yes I had and she mentioned that *that* is an example of what can happen when people talk about animals (Condon, unpublished field notes, October 3, 1992).

In Clyde River (an Inuit community on Baffin Island, Nunavut), elders noted that establishment of a quota in 1973 on killing bears would have a similar effect and cause the bears to go away. When the quota was reduced to zero animals one year, the elders were not surprised (George Wenzel, personal communication). Fienup-Riordan (1983) discusses the importance of maintaining a proper attitude and treating the spirits of the animals appropriately to ensure that the animals will offer their bodies up again during the next seasonal cycle. She also notes the potentially devastating consequences of disrespecting or mistreating animals; once an animal's spirit has been mistreated, it is gone forever and will never return. This point was emphasized by Yup'ik elders when several walruses were found beheaded, their carcasses left to rot on the beach (Fienup-Riordan 1983:xix).

Richard Nelson has written extensively about indigenous attitudes toward the environment, and he has documented the extensive and detailed knowledge that Inupiat (1969), Alaskan Kutchin (1973), and Koyukon (1983) have of their respective environments. This includes knowledge about where to find resources, the best and surest methods of taking them, and the importance of recognizing the dangers of taking more than is needed. Local religious beliefs and attitudes reinforce this knowledge and place it in a larger framework that gives meaning to the universe. Nelson (1982) also points out that the detailed knowledge these people possess about their respective environments should not be surprising. Not only is such knowledge essential for survival, but the Arctic and subarctic ecosystems are considered to be relatively simple, at least from an ecological standpoint. They are also considered to be relatively unproductive. Making assumptions that humans inhabiting these environments are engaged in an obsessive struggle to survive and cannot possibly overexploit their resources, is premature, however. Even with simple technology and small populations, Nelson argues, hunters can very easily affect their environments for the worse.

A related argument is familiar to students of anthropology: Sahlins (1972) has argued that foragers are the original "affluent society," based in part of the absolute amounts of time spent engaged in subsistence. Although foragers are required to lead a nomadic existence and are constrained by a certain degree of environmental scarcity, they have successfully adapted to their environments and are enviable because they focus on meeting their immediate needs and limiting their desires (Sahlins 1972:24). Lee's work with !Kung foragers (1968; 1969) in Botswana, demonstrating that each adult labors in food collection or hunting scarcely fifteen to twenty hours per week, is often cited as support of Sahlins's position.

Sahlins further argues that foragers and other kinship-based societies (horticulturalists in particular) can also be considered "affluent" because they are deliberately underproducing goods, which is itself a form of conservation (Sahlins 1972:49). The natural question that should follow is, of course: Is it true? It is one thing to speak of conservation ethics and discuss the possibility that Native peoples manage or conserve their resources, but it is quite another to demonstrate that such a phenomenon occurs. More recent work on foragers' work effort has questioned Lee's assertions. Hawkes and O'Connell (1981), for example, have shown that Alyawara (a group of central Australian foragers) women spend between four and ten hours per day in gathering and processing food, a figure much higher than that given for !Kung women. Hill et al. (1985) have likewise demonstrated that Ache (a group of foragers in Paraguay) men labored approximately seven hours per day in subsistence. These higher estimates of foragers' work effort can be traced to methodological considerations. Whereas Lee divided subsistence labor into three different spheres (hunting and gathering, tool manufacture and repair, and food processing), later studies have included time spent both processing food and making or repairing tools, which are important components of the food quest, even if they occur in camp. Lee himself has included tool manufacture and repair in revised estimates of work effort. The result is that !Kung work effort is greater than originally calculated—approximately forty-four hours per week for men and forty hours per week for women (Lee 1979:278). The general consensus of anthropologists addressing this issue is that foragers can not and should not be considered "affluent" in the sense that Sahlins uses the term. Foragers spend long hours daily in subsistence tasks and thus do not maximize their leisure time (see also Hames 1992 for a more comprehensive discussion).

Other evidence suggests that not all foragers are deliberately underproducing or underutilizing their resources. Alvard (1993; 1995), for example, has conducted fieldwork among Piro hunters in Amazonian Peru and argued that Piro hunters pursue a hunting strategy in accordance with predictions generated by optimal foraging theory. Alvard defined conservation as the payment of a short-term cost to receive a long-term benefit, and he argues that Piro do not conserve given this definition. Piro do not selectively choose specific age and sex types (such as refraining from taking young females while taking only older males), for example, and he concludes that what has often been called "natural conservation" is in fact simply epiphenomenal and appears only because Piro lack the means to overexploit their resources. This is not to say that conservation cannot appear as a successful adaptive strategy. Among the Piro, there is little if any pressure to practice conservation strategies. Beckerman and Valentine (1996) note that in the blackwater region on northwest Amazonia there exist numerous examples of true conservation. They cite the existence of de facto game reserves, the practice of keeping swidden fields at some distance away from rivers, and the practice of fishermen limiting their catches when fish are spawning. Beckerman and Valentine suggest that these measures are particularly suited to the region, which is well-known as an unproductive environment.

Taken together, the important point is that in many cases foragers are seemingly not conservationists, at least in the sense that Alvard and others have defined conservation (see Hames 1987; 1991). However, the examples cited by Beckerman and Valentine suggest that there are environmental and social conditions that can foster the appearance of true conservation as an adaptive strategy. The following section will examine in greater detail several cases of what appear to be indigenous resource management (and hence conservation) systems in the Canadian Arctic and subarctic.

Cree Systems of Resource Management

The Eastern James Bay Cree provide excellent examples of resource management practices that have remained generally intact over time. Cree are especially noteworthy for the existence of hunting territories allotted to specific individuals, a system with similar incarnations among the neighboring Naskapi-Montagnais (Leacock 1954) and Ojibwa (Bishop 1970). There has been considerable debate about the antiquity of the system, and most anthropologists favor the view that the hunting territory system developed in response to the fur trade. This by itself does not threaten the validity of it as an example of indigenous management, however, for it appears to have arisen locally and has been maintained in the absence of outside authorities. As will be shown below, the hunting territory makes good ecological sense, especially as a management tool for beaver, a resource that is particularly susceptible to overexploitation. The information reported here is taken primarily from the writings of Berkes (1981a; 1987) and Feit (1982; 1988).

The effectiveness of the system relies on the tallyman, or beaver boss, a hunter who is identified with a specific hunting territory and recognized as the territory's "owner." The boss is invested with authority over the land and reserves the right to make decisions regarding the use of the resources on it. He is also endowed with the ability to grant others access to the territory for purposes of hunting big game or trapping furbearers. These associates are typically, but not always, close kin. The boss's authority is not absolute, however. Although the boss and his recognized associates retain rights to big game and furbearers, any hunter may use the resources in the territory on a short-term basis, while in transit or in an emergency.

An additional constraint on the boss's authority is that he is not recognized as the owner of the territory in a traditional capitalistic sense. Berkes writes that:

> The hunters say the land and animals belong to God; the boss is given not really the animals but the *responsibility* to the distribution of the wealth of the land. The boss manages the harvesting activity for the benefit of the band society as a whole: he leads the hunt, supervises the sharing of food, and enforces customary laws with respect to harvesting activities. The boss inherits the hunting territory, usually from his father or an uncle; he cannot sell it or buy it. If he does not manage it for community benefit, he can be forced by social pressure to step down (Berkes 1987:71, emphasis his).

The system is further recognized as an efficient and viable means of effectively managing the resources in a specific hunting territory. Feit writes the following:

> This hunting territory system is an effective means of conserving wildlife, more effective in fact than a system in which hunters were less flexible, because it permits a management of wildlife populations by a single individual, while also permitting a fluid redistribution of hunters to resources. This provides a complex series of strategies for changing the demand for harvest of particular wildlife resource populations in accordance with the changing conditions of the wildlife. The distribution of hunters and dependents can be changed from year to year, territories can go unused so game populations can recover from declines, the "mix" of wildlife resources used to meet subsistence needs can be varied, and the actual level of production of subsistence foods can be modified within certain limits (Feit 1982:387-388).

Feit also cites the extensive knowledge that a boss has of his own territory, especially of beaver lodges, noting that a boss will know the location of each beaver lodge in the territory and usually has a detailed knowledge of the age and sex of the beavers that had been taken from each lodge the last time it had been trapped. This knowledge is used to make future decisions about which lodges to trap and how intensively they should be exploited (Feit 1988).

The Cree also provide another example through fisheries management practices. The Eastern James Bay Cree of Chisasibi (formerly Fort George), Quebec, on eastern James Bay have been studied by Berkes (1977; 1981a; 1985; 1987), who has provided detailed descriptions and analyses of local fisheries practices. Berkes notes that the fishery in the region is used almost exclusively for subsistence, and it is characterized by a high degree of regularity and predictability. Although there are no recognized property rights of access to particular fishing locations, individuals are recognized as belonging to particular locations by their association there over time or, in cases where fishing areas are on a particular boss's land, fishers obtain permission first. The process is largely a simple acknowledgment of authority, however (Berkes 1987). Close to the community (within a nine- to ten-mile radius), fishers use nets with a 2.5 inch mesh, which targets adult-sized cisco. Away from the community (beyond a nine- or ten- mile radius), Cree use three and 3.5 inch nets, which target adult-sized whitefish, a species that is somewhat larger than cisco. Berkes also says that whitefish are relatively scarce within the nine- to ten- mile radius, and he attributes this to the use of smaller mesh sizes. This is also taken as evidence that Cree, whether consciously or unconsciously, have made a decision to focus on one species in one area and another species in another area, the drawback being that whitefish have been overfished close to the community. Overfishing of whitefish is, however, considered to be a minor difficulty because of the large areas away from the community that are lightly fished or not fished at all. Furthermore, the mesh sizes are ideal for catching mature, reproductive fish only and allowing the immature fish to escape. The end result is a productive fishery with high yields, predictable catches, and stability over time (Berkes 1977).

The system works partly because there are numerous cultural institutions that prevent individuals from cheating on the system. The use of smaller mesh sizes is strongly frowned upon and enforced by public opinion, something Berkes noted when he experimented with a number 2 net to see what kinds of fish it caught (Berkes 1977). Other mechanisms include cultural values that stress the importance of not wasting any fish, and social pressure discourages individuals from setting nets too close to others, setting them too deep into the water, or stealing fish from other peoples' nets. There is also a strong self-limiting principle for individuals to take only what they need, which is reinforced by pressure to share the catch with others in the community. The system apparently works so well that Berkes observed few if any rules violations and subsequent punishment (Berkes 1987).

Conditions Where Management Systems Break Down

Berkes has noted (1985) that indigenous management systems are inherently fragile and easily disrupted by any of four factors: 1) commercialization; 2) conversion of controlled access, common property conditions to open access conditions; 3) technology change; and 4) population growth. The well-documented collapses of subarctic furbearers and other game in the 18th and early 19th centuries, for example, has been attributed to the influx of Europeans who neither respected the existing rights of access to the resources nor were interested in sustained fur yields over time. Berkes argues that the Cree and other Algonquians were then faced with a choice. They could either increase fur harvests accordingly and realize some benefit or allow non-Native to take more furs and realize only the costs (Berkes 1981a). The recovery of beaver and the re-emergence of trapping territories in the early part of this century is largely attributed to the Canadian government providing legal recognition of a traditional practice, thus recreating closed access, common property conditions. However, recent development in the area, especially the James Bay Hydroelectric project and associated road construction, has eroded the effectiveness of the system by allowing southern sports hunters access to Cree lands while also facilitating conflicts with hunters from other communities (Berkes 1981b).

The conversion of communal ownership to open access occurring simultaneously with commercialization is not limited to fur trapping across the North American subarctic. This dual process seems to be responsible for management breakdowns and subsequent overexploitation in numerous cases, including the Arctic whale fishery, the Northwest Coast salmon fishery, and the extermination of the Great Plains buffalo herds during the past century (Usher 1987). Technology change and population growth are also implicated in causing these disasters, but it is difficult to determine the roles they play throughout the process. The overfishing of salmon, for example, occurred using the same "simple" technology—fish traps—used by local indigenous peoples (Berkes 1981a). Nevertheless, technology and population growth are implicated even in the absence of commercialization, although Usher (1986:8-9) suggests that evidence supporting such accusations has rarely if ever been conclusive.

It is often assumed, for example, that turn-of-the-century caribou hunting on the Barren Grounds became an environmental disaster because of the introduction of the repeating rifle, which in turn inspired wasteful hunting practices by Natives (see Kelsall 1968:216-225). Overlooked in this argument, however, is the influx of non-Native hunters in the region, the value of caribou skins in the fur market, and a series of severe winters that devastated the caribou populations through starvation, which in turn led to starvation among some Caribou Inuit (Burch 1986). Nevertheless, the assumption that technological change causes resource depletion lives on:

> This use of modern hunting technology has tipped the balance greatly in favor of the Native hunter—the caribou herds are now accessible at virtually all times of the year. The so-called harmony between primitive Native hunters and caribou was imposed by the caribou's continuous movements; the Native's relative lack of mobility; and the Native's poor weaponry. Today, aircraft pilots and hunters using two-way radios relate the locations and movements of

> the caribou during their migrations and wintering period, wherever outpost camps exist. The caribou now have little or no chance of passing camps undetected, when the hunters have been alerted to their coming by radio and are searching the countryside by snowmobile. Once the caribou are found, snowmobile-mounted hunters can work the groups by herding them about until the hunters have taken as many as they want—which in the case of relatively small groups, is all the caribou (Miller 1983:173).

As with the tragedy of the commons argument, arguments about technological change frequently overlook the social and cultural context in which the technology appears. William Kemp, for example, has documented the flow of energy in a Baffin Island community and notes that snowmobiles, rifles, outboards, and other mechanized equipment provide certain advantages over manual foraging. What is often overlooked is that he notes the necessity of this technology for effective foraging from newly centralized communities, so that in many ways the use of the technology is a trade-off. He also recognizes the social constraints on acquiring and using that technology, although he does not elaborate: "At the same time that motorized transport has enhanced the ability to kill game, other social and economic factors have acted to reduce the amount of time available for hunting and have kept the kill within bounds" (Kemp 1971:115).

A somewhat more detailed analysis comes from the community of Holman. During 1992-1993, Richard Condon and I conducted research with a sample of young adult household heads. We were interested in documenting the effects of social changes on members of our sample, particularly with regard to the economic decisions they made about engaging in subsistence hunting, wage labor, or a combination of the two (specific methods and results can be found in Condon et al. 1995). During the course of fieldwork, it became clear that engaging in hunting and trapping activities were important components of a young man's sense of identity as both an adult and an Inuk. To be recognized as a true Inuk required being a "doer," one who hunted and trapped and provided food for others, but many of the members of our sample were prevented from doing these things by a combination of factors, including 1) lack of skill and knowledge necessary for hunting and trapping, which we attributed to community centralization and the collapse of the fur markets in the early 1980s; 2) lack of capital to afford the high costs of purchasing and maintaining equipment, and 3) time constraints arising from work obligations for those who could afford to hunt. Indeed, although the most active hunters in our sample were also wage earners with high-paying jobs, it was clear that working a job severely limited their opportunities to hunt. The two most prolific hunters in our sample, for example, were virtually stuck in the settlement for two months because of bad weather on the weekends.

There is also no indication that new technology has led to resource depletion through wasteful practices or overexploitation. Damas (1972; see also Wenzel 1995) has noted that there have been few if any detailed analyses of exchange systems among Inuit groups, and this fact raises an important issue. Examinations of subsistence hunting and resource use have almost invariably focused on production while ignoring the social organization of the community. One might argue that a hunter could conceivably hunt more than he could possibly use, but an important limiting constraint is that even if more was taken than was necessary, the excess will invariably be shared with other households. Indeed, it is often expected that this will be the case, and a hunter may never have the opportunity to offer meat to others simply because others come and help themselves. The sharing network, in addition to its functioning as a mechanism for distributing resources through the community, also functions as an information network. Although sharing of excess is expected and highly regarded, there seems to be a limit to how much excess a hunter can harvest. Hunters who do harvest more than can be shared or consumed in a reasonable period of time are tolerated but publicly stigmatized.

Holman duck hunting is an excellent example of both social pressure to conform to reasonable expectations of use and the self-limiting principle. Holman is fortuitously situated along the migratory route of king eider ducks, and each June flocks fly down the coast. The importance of the hunt is so great to community members that town nearly empties during this month as families move to camps up and down the coast for several miles. A particularly good spot is known as Mashuyok, located some five miles from the settlement at the narrow channel that separates Victoria Island from Holman Island. Many people congregate there to await the migrating flocks, and as they pass overhead people shoot at them with shotguns. This is an important time of year for Holman residents, especially because it is close to the settlement and even those with little capital can usually afford to participate in the hunt.

Ducks are a tasty and easily harvested food source, and also provide an important link to Inuit identity. For some young men, this may be the only hunting they do all year. Even though the ducks are susceptible to wasteful hunting practices by taking much more than is needed, this is almost unheard of. The few cases of hunters throwing the previous year's leftover ducks in the dump to make freezer space for new occupants were met by stern public disapproval. During our interviews with informants, we asked about duck-hunting efforts and their target harvest. All of our informants reported that they shot x number of ducks and geese that year, which they felt was enough to both support their family's current craving for fresh waterfowl and provide some variability in the diet during the winter. Most claimed that they based their hunting effort on their consumption during the previous year, either adjusting upward several ducks and geese because they ran out the year before or taking a few less because there were some left in the freezer when the new season started. A number of elders indicated that they no longer even picked up a shotgun during duck season because their grandchildren hunted ducks that the elders knew would end up in their pot or freezer. They also effectively monitored their grandchildren's efforts because they were providing them with shotguns and ammunition.

The self-limiting behavior of the duck hunters is both ethical and practical. People frown upon wasting meat, even ubiquitous foods such as ducks, but there is also a very practical reason for refraining. During the summer, freezer space is limited, and most people prefer to save space in the freezer for fish caught during summer net fishing and for caribou acquired during the summer caribou hunt. Informants also reported that they monitored their own efforts from year to year partly because they did not want to be caught in the dilemma of having leftover ducks in the freezer the following spring. Wasting is unethical and people try to avoid it, but they also try to avoid having to eat ten month-old, freezer-burned foods when the prospect of fresh foods is close at hand.

The biggest wastage during duck-hunting season is, ironically, of ammunition. The recent preponderance of pump shotguns in Holman has, according to elders, led to an increase in ammunition costs during duck season because pump shotguns, unlike manually loaded single- or double-barreled shotguns, lend themselves to shooting indiscriminately rather than taking deliberate aim. The "blast away" method was favored primarily by younger hunters, but many noted that it was much less efficient than a more conservative approach, especially in terms of the costs of shotgun shells.

The Holman example is of course anecdotal, and neither Condon nor I collected data for the purposes of a detailed cost-benefit analysis of the hunt. A more sophisticated look at costs and benefits of foraging can be found in Eric Smith's (1991) work with Inujjuamiut hunters on the east coast of Hudson Bay. Based on his analysis of mechanized and manual foraging, Smith argues that mechanized foraging is more costly than relying on alternate economic strategies and consuming imported food. Explanations of why foraging persists in spite of the greater costs are problematic, however. Explanations include 1) the important social and psychological benefits derived from hunting (Condon et al. 1995; see also O'Neil 1984); 2) speculation that the observed equilibrium between foraging and wage labor in the Arctic is really a gradual transition from a foraging economy to a cash and import economy, with cultural values further slowing the process or even preserving the balance (Smith 1991:394; see also Feit 1982); and 3) that country food is considered to be culturally superior to imported food (see Condon et al. 1995; Freeman 1988a; Wein et al. 1996) or nutritionally superior to imported food, and thus worth the greater cost (Freeman 1988b; Freeman et al. 1992; Usher 1976).

Concerns about population growth fueling overexploitation of resources are equally important but analyzed with even less frequency. Although it is assumed that the rapid growth rates of many northern communities will severely tax local wildlife resources, it is important to note that the entire population of the Northwest Territories is only slightly greater than it was one hundred years ago (Usher 1986:8-9). Furthermore, the advent of mechanized transport has reduced the number of dog teams kept by hunters, thus reducing demands on wildlife resources as dog food. Additionally, consumption of country foods is not as great as it was a century ago. In Holman, Condon and I recorded food consumption with our sample and found that, on average, roughly half of a household's meals consisted of country food, the other half being store foods imported from the south (see Condon et al. 1995:43). Although we were unable to measure exact amounts consumed or calculate caloric intake, our estimate agrees with other reports of food consumption (see Feit 1982:382).

In summary, claims that technological change and population growth fuel resource depletion because they upset the balance of a predator/prey equilibrium appear to be somewhat exaggerated. While these factors may influence resource depletion, and they may become important factors in the future, especially if northern populations continue to grow, at present it seems that social and cultural limitations are effective mechanisms for preventing these factors from appearing as significant influences on resource overexploitation.

The Political and Social Contexts of Wildlife Management

The Culture of Wildlife Management

To this point this paper has examined the merits of a basic of assumption held by wildlife managers—that local people cannot manage their own resources, especially in the face of rapid and dramatic social change. I have argued that this assumption suffers from a number of problems and overlooks the important social and cultural arrangements that affect local management regimes and resource use patterns. In spite of increasing attacks by anthropologists and others, however, these beliefs persist, perhaps because of a more general and firmly held belief that wildlife management is a scientific endeavor, and science is perceived as a superior way of knowing about the world.

Milton Freeman has written extensively about the culture of wildlife management, and he argues (1985) that one of the validating principles of wildlife science is quantification. Quantification is seen not only as a means of attaching values to observations and perceptions but also the only means of understanding environmental phenomena. Quantification becomes the measure by which all other investigative endeavors are judged, and the absence of quantification is taken as a signal that the research is suspect and the conclusions questionable because the objectivity of the researcher has been compromised and the conclusions cannot be said to be decisively true.

Science is therefore seen as an objective undertaking. Rigorous adherence to quantification, modeling, and hypothesis testing will prevent potential abuses to the system. Practices such as peer review of publishable materials is also thought to back up the integrity of the system and allow for detection of fraud. While this belief is firmly entrenched both within the scientific community and among the general public, Freeman (1989) cites numerous ways in which individuals promote their careers at the expense of science. Notable strategies include publishing trivia, publishing in obscure venues, falsifying data, and withholding raw data from other researchers who request access to it.

Given the intense weight granted to "scientific" explanations of phenomena, it should not be surprising that many individual scientists will engage in public speculation or speak out on topics they feel strongly about. Speculation is not necessarily a bad thing, but there is a reduced burden of proof outside of the professional community, and "scientists" may be tempted to engage in debates that may not be trained to understand. Freeman writes:

> This indulgence in "trans-science" is by no means uncommon when scientists assume an advocacy stance on a popular or political issue. Given the public attention that the mass media can accord to scientists speaking out on topical issues, it may take little more than subsequent citing of a trans-science observation in a mass-circulation magazine or newspaper article for a scientifically unsubstantiated opinion to become uncritically accepted by large numbers of people as an established, "scientific," fact (Freeman 1994:144).

In summary, wildlife management like other scientific endeavors exists within its own unique social context, of which managers are frequently unaware (Osherenko and Young 1989:118). Freeman demonstrates the affect that social and political context can have on a scientific undertaking in a 1989 paper in which he examines an assertion made by wildlife biologists that Canada was in the midst of a caribou crisis. In a federal government report published in 1985, wildlife managers noted with alarm that during the period from 1970-1980 most of the caribou herds in Canada were in sharp decline. This crisis was perceived to be continuing into the 1980s. According to Freeman, the problem with the report was that subsequent analysis of the available data showed that some of the herds reported to be decreasing were in fact increasing. He attributes the misinterpretation of the data to multiple causes, including the possibilities that: 1) managers used data that was either incomplete or collected under questionable circumstances, but they neglected to mention these shortcomings; 2) managers excluded more recently published data that was more thorough and showed an increasing, rather than decreasing, trend; 3) managers consciously or unconsciously manipulated the data to support their claims; 4) managers were reluctant to share management responsibilities with Native groups; and 5) Environment Canada was facing budget cuts and the loss of over 400 positions. Taken together, Freeman argues that the caribou crisis did not exist, but a wildlife management crisis did, and managers wanted to convince the government that financial support for research and administration was still necessary.

Freeman's account makes it clear that wildlife management is limited as a scientific endeavor because its cultural and political context compromises the objectivity of those engaging in it. As noted above, the primary goal of wildlife management is to develop scientific knowledge of wildlife for the purpose of sustainable human use. This is a noble goal but one fraught with problems in the Canadian Arctic and subarctic. One of the biggest problems is the incomplete knowledge wildlife managers have of almost all managed species in the region. Feit (1988), in an analysis of beaver management in eastern James Bay, notes that wildlife managers seeking to manage local beaver populations had to rely on published information on beavers studied in Michigan and Alaska. After more intensive study of the James Bay beaver populations, however, it became clear that management policies had to be changed because their predictions were incorrect.

In addition to a limited body of knowledge about certain species, there is also a limited time depth to observations. Burch (1972) has noted that caribou populations are generally thought to fluctuate in cycles ranging from thirty-five to one hundred years, which is a time frame far greater than that covered by observations made by wildlife managers. In the Holman region, for example, the local caribou population had never been surveyed before 1978 (Miller 1978). In the absence of a considerable time frame, it becomes difficult to gauge the affect of such practices as human hunting on the population.

An additional concern is the focus of wildlife management. Unlike ecology, which seeks to understand the operation of entire ecosystems and the nature of complex interactions between species, wildlife management focuses on only one species at a time. To again use an example from the Holman region, during the 1992-1993 caribou crisis Inuit hunters offered several explanations why caribou were scarce. One explanation was that caribou were scarce in response to the steady increase in muskoxen populations over the previous decade (following their reintroduction to the region and a ban on hunting them). They argued that caribou and muskoxen dislike each other and the caribou simply moved away. Inuit did, however, recognize that the two species prefer different niches, and they do not generally compete for food or space, an assertion supported by a number of different researchers (Gray 1984:140; Gunn 1984; Kevan 1974; Miller et al. 1977; Vincent and Gunn 1981). Wildlife biologists admit, however, that the relationships between the two species are poorly understood (Gunn 1984), and there may be many other factors involved in the interaction between the two species. Some Inuit hypothesized that increases in the muskoxen population had been followed by increases in the wolf population, and although muskoxen is a predictable and secure resource for wolves, wolves prefer to eat the more difficult to capture caribou (see Crissler 1956), which in turn may encourage caribou to move to new areas reasonably devoid of muskoxen. Other possibilities are that Victoria Island has conditions that facilitate niche overlap or outright competition between the two species, a phenomenon observed in other regions of the Arctic (Klein and Staaland 1984; Thomas and Edmonds 1984). Currently, the answers to these questions are as yet unknown, and one would expect that detailed knowledge of the relationships between species should be obtained before management decisions are made.

A final consideration is that wildlife biology strives to be a scientific exercise, but wildlife management is not. Peter Usher argues (1986:92-100), for example, that while management's goal is to manage resources effectively for sustainable human use, there are numerous different groups of humans seeking to use a resource, each of which has a different agenda. Managers therefore have the unpleasant task of making decisions about appropriate and sustainable use within the constraints set by competing users, which can include Native groups, sports hunters, industry, the state, and the general public, to name just a few. Usher writes:

> Although science and the scientific method are clearly useful in estimating, for example, the numbers of animals available for harvest on a sustained-yield basis, and for projecting the effects of human activity on these numbers, *allocation* of the resource has very little to do with the biological

> consequences. Allocation, which is a central issue in wildlife management in the North, is largely a matter of balancing equity and efficiency—in other words, an economic and political exercise (Usher 1986:95).

Wildlife managers can also become entangled in the politics of management when they begin to see themselves as one of the few groups advocating the rights of the animals themselves. This in turn may lead managers to take a proprietary view of their study populations, fostering the conviction that only managers know what is the best management policy. Furthermore, managers may come to realize that their effort is a losing cause in the face of economic growth and development in the North (Theberge 1981). Additional pressure is placed on managers by Native groups, who have a genuine concern for the use of specific resources for their own economic well-being. Indeed, this struggle can quickly become more than a concern for economic well-being; it is an expression of concerns about political autonomy (Berkes et al. 1991; Nakashima 1993).

The Utility of Traditional Ecological Knowledge

In the previous section it was noted that local indigenous knowledge of the environment is regarded as inferior because it is not quantitative. It is additionally scorned because it is frequently considered to be no more than anecdotal, and compromised by spiritual beliefs that are often regarded as little more than superstitions. Freeman, however, cautions against dismissing indigenous knowledge as inferior and useless because both science and indigenous forms of ecological knowledge are built upon the same foundations:

> [A] reason for not treating local knowledge lightly is because it rests upon the same foundations as all modern science, namely empirical observation and deductive logic. The very foundations of the biological sciences are based upon no more than that, for the works of Charles Darwin and Alfred Russell Wallace, though epochal in magnitude, were no more than those of keen natural historians armed with magnifying glasses, tweezers, and pinning boards. All of modern biology rests upon a taxonomic system elaborated a century before Darwin and Wallace made their discoveries; today thousands of plant and animal species still bear the Linnean designations that 18th century Swedish naturalist and his students determined to be those species' true place in the natural order. Coming closer to modern times, we can recall that the trophic-dynamic understanding of ecosystems was elucidated by a youthful Charles Elton about 60 years ago, following a field season contemplating the Svalbard tundra community, at which time he was also equipped with little more that binoculars, a vasculum, and a rifle. On what grounds, therefore, can

> we doubt the utility of the knowledge of local people derived from many more years of patient observation in an environment where their daily well-being, if not their physical survival, requires a thorough understanding of how nature works? (Freeman 1987:70-71).

Resistance to the possibility of using oral tradition to inform scientific inquiry is perhaps understandable because oral tradition contains two different kinds of knowledge. On the one hand, it contains knowledge about the environment that is inherently useful for survival. On the other hand, however, traditional ecological knowledge is imbued with detailed explanations of the nature of human and environmental relationships, explanations which are eminently useful to indigenous peoples as a means of instructing others in how the world operates and how people should behave as members of that particular society. The incorporation of two kinds of knowledge in one format should not discourage those from attempting to learn something from it. Julie Cruikshank (1981), for example, has examined various spheres of knowledge among southern Yukon Native peoples and detailed the convergence of local knowledge with Western, scientific explanations of phenomena ranging from knowledge about glacial movements, climate changes, and local flora and fauna. She also argues that traditional forms of ecological knowledge are useful not only for the survival of the people relying on it but also for the formation of testable hypotheses.

Another use of traditional ecological knowledge is in its ability to inform researchers about aspects of the environment or a particular species' behavior that would otherwise remain unobserved. Douglas Nakashima (1993) has documented the extensive knowledge that local people have of common eiders living along the southeast coast of Hudson Bay. This knowledge includes an extensive natural history of the winter ecology of common eiders, including the winter distribution and survival strategies of the birds, and the relationships between local environmental conditions and survivorship. The information collected from Inuit is valuable primarily because it is natural history about a species and its ecology that most ecologists and wildlife biologists would never obtain, simply because wildlife professionals are rarely in the field during the period in question (Johannes 1993).

In the contemporary North, it is becoming increasingly common for managers and Native groups to come together in what is called co-management, where expertise from both parties is combined and utilized in making decisions about the use of wildlife resources. The idea is that the combined knowledge of both groups is greater than the knowledge of either group alone; this allows for equal participation in making the often difficult political and economic decisions regarding resource use. Berkes (1981a) notes that there are generally three different arrangements that are considered co-management. One such arrangement involves including Natives as participants in scientific studies or in cooperative research projects as a means of developing a northern Native scientific community. A second approach relies extensively on southern science to set the framework and guidelines for management and includes local groups in management decisions. Both of these paradigms are not ideal, however, because they exclude Native peoples from part of the process—either the political process, as in the first arrangement, or the research process, assuming that Natives have little knowledge to offer, as in the second arrangement.

Berkes goes on to say that a third arrangement is the ideal one for which to strive, and it involves including Natives both in the research process and in making management decisions. He notes that this last approach not only ensures equal participation and provides a framework for political autonomy, but it can be quite successful when implemented. As an example, he cites the Cree among whom he has worked (Berkes et al. 1991) and notes that during the mid-1980s, local trappers had noticed a decline in the beaver populations in the region. This led Cree to conduct a detailed inventory of the beaver lodges in the area, and the community decided to close beaver trapping in areas where beaver populations were low. The ban was lifted several years later when a restudy showed the populations had recovered sufficiently for trapping to begin again. Berkes argues that the problem was not only detected at the local level, but it was resolved at the local level with minimal intrusion by outside agencies.

In spite of demonstrated success in co-managing resources and emphasizing management at the local rather than regional level, wildlife managers continue to resist incorporating Native viewpoints in their research. Indeed, when managers profess their beliefs in the importance of co-management, they invariably suggest that the key is to train more Natives in the goals of science, yet they refrain from suggesting that wildlife managers should learn more about Native knowledge systems (Nakashima 1993). Ian Stirling, a wildlife biologist, writes the following:

> Over the longer term, it is essential that a significant number of Native people become qualified in scientific aspects of wildlife conservation. Native people, with considerable justification, already regard themselves as the experts on wildlife, especially in areas where they have lived and hunted for generations. The knowledge and traditions developed in the past must not be lost, but times have changed sufficiently that it is important now to augment that background. To avoid the suggestion of tokenism in hiring, Native biologists must eventually obtain the same scientific training as everyone else. Initially at least,

> this will be difficult because of having to leave family and familiar surroundings to attend southern universities (Stirling 1990:iii).

Part of the belief in the correctness of this approach to co-management may be that managers and other southerners look at the rapidly changing social environment of the North and assume that social changes are eroding whatever traditional knowledge and management systems once existed, and some form of knowledge must be offered to fill in the void. At the same time, however, this is an insidious process because it undermines traditional systems and is an attempt, whether conscious or not, to convert Natives to a new cultural framework. At the beginning of this paper, I noted that the wildlife manager studying the caribou crisis hired a local Inuk, who was responsible for interviewing hunters about their knowledge of caribou movements and population cycles. This Inuk, however, had been educated during his youth in southern boarding schools and trained in the principles of wildlife management. It became clear during numerous social visits with him that he sympathized with the manager's concerns and believed that local hunting practices were indeed responsible for caribou depletion, even though other elders in the community disagreed with him. Because he was the hired assistant, however, he had the opportunity to report his ideas to the manager and potentially exclude other views, thus hindering community participation in the research.

Conclusions

This paper has primarily focused on reviewing research that has called the beliefs and practices of wildlife management into question while pointing out some hurdles that both managers and Native groups face when they attempt to engage in what is called co-management. It does not advocate, however, that attempts to promote and engage in co-management are fruitless and not worthwhile. On the contrary, this is an exciting time in the Canadian North precisely because it seems that resistance to the idea of co-management is slowly breaking down. The caribou crisis in Holman, for example, was a less-than-heartening experience for many Inuit because they felt they were being blamed for the caribou decline and they believed that had not been involved in management's response to the degree that they should have been. However, the local and regional response to the crisis was generally positive in that Inuit from the neighboring communities of Cambridge Bay and Coppermine expressed their concern for Holman residents and offered to aid their brethren. The Coppermine Hunters' and Trappers' Association, for example, sponsored several caribou hunts in Coppermine in which Coppermine Inuit hunters harvested caribou that was subsequently shipped to Holman and distributed free to elders in the community. A second episode involved shipping caribou meat to the Holman HTA, where it was sold at cost to local Inuit. Holman Inuit also began receiving more shipments of caribou meat via air freight from relatives in the neighboring communities. In the long run, then, the potential damage done to the local economy and the well-being of individual hunters may be offset by increased social contacts and interdependence between communities. Likewise, the Holman region is by all accounts abundant with game, including muskoxen, ringed seals, fish, and many species of migratory waterfowl. Although caribou were scarce, local Inuit were not without country food alternatives.

Nevertheless, caution is necessary. The wildlife manager dealing with the Holman caribou crisis may have left the community thinking that they had just witnessed a successful example of co-management, although the process itself left much to be desired. It is still unclear whether the manager was correct in justifying a hunting ban by assuming hunting activity was the cause. The situation seems to be improving in the North, but the problems outlined in this paper still exist and may still hinder efforts at true co-management in the future.

References

Alvard, Michael S. 1993. Testing the 'ecologically noble savage' hypothesis: Interspecific prey choice by Piro hunters of Amazon Peru. *Human Ecology* 21(4):355-387.

———. 1995. Intraspecific prey choice by Amazonian hunters. *Current Anthropology* 36(5):789-818.

Beckerman, Stephen, and Paul Valentine. 1996. On Native American conservation and the tragedy of the commons. *Current Anthropology* 37(4):659-661.

Berkes, Fikret. 1977. Fishery resource use in a subarctic Indian community. *Human Ecology* 5(4):289-307.

———. 1981a. The role of self regulation in living resources management in the north. In *Proceedings: First international symposium on renewable resources and the economy of the north, Banff, Alberta, May 1981*. M.M.R. Freeman, ed., 166-177. Ottawa: Association of Canadian Universities for Northern Studies.

———. 1981b. Some environmental and social impacts of the James Bay hydroelectric project, Canada. *Journal of Environmental Management* 12:157-172.

———. 1985. Fishermen and the tragedy of the commons. *Environmental Conservation* 12(3):199-206.

———. 1987. Common-property resource management and Cree Indian fisheries in subarctic Canada. In *The question of the commons: The culture and ecology of communal resources.* B. J. McCay and J. Acheson, eds., 66-91. Tucson: University of Arizona Press.

Berkes, Firket, and D. Feeny. 1990. Paradigms lost: Changing views on the use of common property resources. *Alternatives* 17(2):48-55.

Berkes, Firket, Peter George, and Richard J. Preston. 1991. Co-management: The evolution in theory and practice of the joint administration of living resources. *Alternatives* 18(2):12-18.

Bishop, Charles A. 1970. The emergence of hunting territories among the northern Ojibwa. *Ethnology* 9(1):1-15.

Burch, Ernest S. Jr. 1972. The caribou/wild reindeer as a human resource. *American Antiquity* 37(3):339-368.

———. 1986. The Caribou Inuit. In *Native peoples: The Canadian experience.* R.B. Morrison and C.R. Wilson, eds., 107-133. Toronto: The Canadian Publishers.

Condon, Richard G., Peter Collings, and George Wenzel. 1995. The best part of life: Subsistence hunting, ethnicity, and economic adaptation among young Inuit males. *Arctic* 48(1):31-46.

Crissler, Lois. 1956. Observations of wolves hunting caribou. *Journal of Mammology* 37(3):337-346.

Cruikshank, Julie. 1981. Legend and landscape: Convergence of oral and scientific traditions in the Yukon Territory. *Arctic Anthropology* 28(2):67-93.

Damas, David. 1972. Central Eskimo systems of food sharing. *Ethnology* 11(3):220-240.

Feeny, David., Fikret Berkes, Bonnie J. McCay, and James M. Acheson. 1990. The "Tragedy of the Commons": Twenty-two years later. *Human Ecology* 18(1):1-19.

Feit, Harvey A. 1982. The future of hunters within nation-states: Anthropology and the James Bay Cree. In *Politics and history in band societies.* E. Leacock and R. Lee, eds., 373-411. London: Cambridge University Press.

———. 1988. Self-management and state management: Forms of knowing and managing northern wildlife. In *Traditional knowledge and renewable resource management.* M.M.R. Freeman and L.N. Carbyn, eds., 72-91. Edmonton: Boreal Institute of Northern Studies, University of Alberta.

Fienup-Riordan, Ann. 1983. *The Nelson Island Eskimo: Social structure and ritual distribution.* Anchorage: Alaska-Pacific University Press.

Fratkin, Elliot. 1991. *Surviving drought and development.* Boulder, CO: Westview Press.

Freeman, Milton M.R. 1985. Appeal to tradition: Different perspectives on Arctic wildlife management. In *Native Power*. J. Brøstad, J. Dahl, A. Gray, H.C. Gulløv, G. Henriksen, J.B. Jørgenson, and I. Kleivan, eds., 265-281. Bergen: Universiteitsforlaget AS.

———. 1987. Concluding remarks: Dissent, diversity, and biosphere reserves. In *Proceedings of the symposium on research and monitoring in biosphere reserves*. N. Simmons, M.M.R. Freeman, and J. Inlgis, eds., 69-72. Occasional Paper #20, Boreal Institute for Northern Studies, Edmonton, Alberta.

———. 1988a. Tradition and change: Problems and persistence in the Inuit diet. In *Coping with Uncertainty in Food Supply*. I. de Garine and G.A. Harrison, eds., 150-169. Oxford: Clarendon Press.

———. 1988b. Environment, society, and health: Quality of life issues in the contemporary north. *Arctic Medical Research* 47 (suppl. 1):53-59.

———. 1989. Graphs and gaffs: A cautionary tale in the common-property resources debate. In *Common property resources: Ecology and community-based sustainable development*. F. Berkes, ed., 92-109. London: Belhaven Press.

———. 1994. Science and trans-science in the whaling debate. In *Elephants and whales: Resources for whom?* M.M.R. Freeman and U.P. Kreuter, eds., 143-158. Basel, Switzerland: Gordon and Breach Science Publishers.

Freeman, M.M.R., Eleanor Wein, and Darren E. Keith. 1992. *Recovering rights: Bowhead whales and Inuvialuit subsistence in the western Canadian Arctic*. Edmonton, Alberta: Canadian Circumpolar Institute.

Gordon, H.S. 1954. The economic theory of a common-property resource: The fishery. *Journal of Political Economy* 62:124-142.

Gray, David R. 1987. *The muskoxen of Polar Bear Pass*. Markham, Ontario: Fitzhenry and Whiteside.

Gunn, Anne. 1984. Aspects of the management of muskoxen in the Northwest Territories. In *Proceedings of the first international muskox symposium*. D.R. Klein, R.G. White, and S. Keller, 33-40. Biological Papers of the University of Alaska, Special Report #4.

Hames, Raymond. 1987. Game conservation or efficient hunting? In *The question of the commons: The culture and ecology of communal resources*. B.J. McCay and J. Acheson, eds., 92-107. Tucson: University of Arizona Press.

———. 1991. Wildlife conservation in tribal societies. In *Biodiversity: Culture, conservation, and ecodevelopment*. M.L. Oldfield and J.B. Alcorn, eds., 172-199. Boulder, Colo: Westview Press.

———. 1992. Time allocation. In *Evolutionary ecology and human behavior*. E.A. Smith and B. Winterhalder, eds. 203-236. Hawthorne, New York: Aldine de Gruyter.

Hardin, Garrett. 1968. The tragedy of the commons. *Science* 162:1243-48.

Hawkes, Kristen, and James F. O'Connell. 1981. Affluent hunters? Some comments in light of the Alyawara case. *American Anthropologist* 83:622-626.

Hill, Kim, Hillard Kaplan, Kristen Hawkes, and Ana M. Hurtado. 1985. Men's time allocation to subsistence work among the Ache of eastern Paraguay. *Human Ecology* 13(1):29-47.

Johannes, R.E. 1993. Integrating traditional ecological knowledge and management with environmental impact assessment. In *Traditional ecological knowledge: Concepts and cases*. J.T. Inglis, ed., 33-39. Ottawa: International Program on Traditional Ecological Knowledge and International Development Research Centre.

Kelsall, John P. 1968. *The migratory Barren Ground caribou of Canada*. Ottawa: Department of Indian Affairs and Northern Development.

Kevan P.G. 1974. Peary caribou and muskoxen on Banks Island. *Arctic* 27(4):256-264.

Kemp, William B. 1971. The flow of energy in a hunting society. *Scientific American* 225:105-116.

Klein, David R. and Hans Staaland. 1984. Extinction of Svalbard muskoxen through competitive exclusion: An hypothesis. In *Proceedings of the First International Muskox Symposium*. D.R. Klein, R.G. White, and S. Keller, 26-31. Biological Papers of the University of Alaska, Special Report #4.

Leacock, Eleanor. 1954. *The Montagnais hunting territory and the fur trade*. American Anthropological Association Memoir 78.

Lee, Richard B. 1968. What hunters do for a living, or, How to make out on scarce resources. In *Man the hunter*. R. Lee and I. DeVore, eds., 30-48. Chicago: Aldine Press.

———. 1969. !Kung bushman subsistence: An input-output analysis. In *Environmental and cultural behavior*. A.P. Vayda, ed., 47-49. Garden City, New York: Natural History Press.

———. 1979. *The !Kung San: Men, women, and work in a foraging society*. Cambridge: Cambridge University Press.

MacPherson, A.H. 1981. Commentary: Wildlife conservation and Canada's north. *Arctic* 34(2):103-107.

McCay, Bonnie J. 1987. The culture of the commoners: Historical observations on old and new world fisheries. In *The question of the commons: The culture and ecology of communal resources*. B.J.

McCay and J. Acheson, eds., 195-216. Tucson: University of Arizona Press.

McCay, Bonnie J., and James Acheson, eds. 1987. *The question of the commons: The culture and ecology of communal resources.* Tucson: University of Arizona Press.

Miller, Frank L. 1978. Numbers and distribution of Peary caribou on the Arctic islands of Canada. In *Parameters of caribou population ecology in Alaska: Proceedings of a symposium and workshop.* D.R. Klein and R.G. White, eds., 16-19. Biological Papers of the University of Alaska, Special Report #3.

———. 1983. Restricted caribou harvest or welfare: Northern Native's dilemma. *Acta Zoologica Fennica* 175:171-175.

Miller, F. L., R. H. Russell, and A. Gunn. 1977. Distribution, movements, and numbers of Peary caribou and muskoxen on the western Queen Elizabeth Islands, Northwest Territories, 1972-1974. *Canadian Wildlife Service Report Series #40.*

Nakashima, Douglas J. 1993. Astute observers on the sea ice edge: Inuit knowledge as a basis for Arctic co-management. In *Traditional ecological knowledge: Concepts and cases.* J.T. Inglis, ed., 99-110. Ottawa: International Program on Traditional Ecological Knowledge and International Development Research Centre.

Nelson, Richard K. 1969. *Hunters of the northern ice.* Chicago: University of Chicago Press.

———. 1973. *Hunters of the northern forest.* Chicago: University of Chicago Press.

———. 1982. A conservation ethic and environment: The Koyukon of Alaska. In *Resource managers: North American and Australian hunter-gatherers.* N.M. Williams and E.S. Hunn, 211-228. AAAS Selected Symposium 67. Boulder, Colo: Westview Press.

———. 1983. *Make prayers to the raven.* Chicago: University of Chicago Press.

O'Neil, John. 1984. *Is it cool to be an Eskimo? A study of stress, identity, coping, and health among Canadian Inuit young adult men.* Ph.D. Thesis, University of California, San Francisco.

Osherenko, Gail, and Oran R. Young. 1989. *The age of the Arctic: Hot conflicts and cold realities.* Cambridge: Cambridge University Press.

Sahlins, Marshall. 1972. *Stone age economics.* Chicago: Aldine-Atherton.

Smith Eric A. 1991. *Inujjuamiut foraging strategies.* New York: Aldine de Gruyter.

Stefánsson, Vilhjalmur. 1913. *My life with the Eskimo.* New York: Macmillan.

Stirling, Ian. 1990. Guest editorial: The future of wildlife management in the NWT. *Arctic* 43:iii-iv.

Swaney, James A. 1990. Common property, reciprocity, and community. *Journal of Economic Issues* 24(2):451-463.

Theberge, John B. 1981. Commentary. Conservation in the north: An ecological perspective. *Arctic* 34(4):281-285.

Thomas, Donald C., and Janet E. Edmonds. 1983. Rumen contents and habitat selection of Peary caribou in winter, Canadian Arctic archipelago. *Arctic and Alpine Research* 15:97-105.

Usher, Peter J. 1976. Evaluating country food in the northern Native economy. *Arctic* 29(2):105-120.

———. 1986. *The devolution of wildlife management and the prospects for wildlife conservation in the Northwest Territories*. Ottawa: Canadian Arctic Resources Committee.

———. 1987. Indigenous management systems and the conservation of wildlife in the Canadian north. *Alternatives* 14(1) :3-9.

Vincent, D. and Anne Gunn. 1981. Population increase of muskoxen on Banks Island and implications for competition with Peary caribou. *Arctic* 34:175-179.

Wein, Eleanor E., M.M.R. Freeman, and Jeanette C. Markus. 1996. Use of and preference for traditional foods among the Belcher Island Inuit. *Arctic* 49(3):256-264.

Wenzel, George. 1995. Ningiqtuq: Resource sharing and generalized reciprocity in Clyde River, Nunavut. *Arctic Anthropology* 32(2):43-60.

3

"Aboriginal Nations": Natives in Northwest Siberia and Northern Alberta

Aileen A. Espiritu

Tuberculosis, alcoholism, suicide, high infant mortality, malnutrition, cultural and linguistic decimation, racial discrimination, destruction of Native lands, loss of control over social and economic organization—the list is endless. These characteristics and experiences are shared by aboriginal peoples of two very disparate polities—Russia and Canada. It is quite remarkable that the plight of Native peoples of Russia and Canada should be so strikingly similar. A comparative study may be instructive in exploring "critical commonalities," as Stephen Cornell suggests in his study of American Natives (1988:vi).

Within the framework of aboriginal-state relations posited by Augie Fleras and Jean Leonard Elliot, the Khanty, Mansi and Yamalo-Nenets of northwestern Siberia and the Lubicon Lake Indian Band of Alberta may be analyzed as "nations within." Quite simply,

> aboriginal people assume the politically self-conscious stance of 'nation' when they go a step beyond identifying themselves as the 'original' occupants of a

I would like to thank Dr. Eric Smith for his insights and advice regarding this chapter, and both he and Joan McCarter for their careful editorial work and sustained patience. My thanks also goes to Dr. Kurt Engelmann and REECAS for providing the maps, and for the invitation to the symposium at which this paper was originally presented.

Research in Northwest Siberia in the summers of 1993 and 1994 was supported by the Government of Alberta Department of Education, the Canadian Circumpolar Institute, the Association of Universities and Colleges of Canada Professional Partnerships Programmes and the University of Alberta Department of History. None of these organizations is responsible for the views expressed.

> land who wish to preserve and protect their cultural heritage. This additional step consists in their assertion that they have a special relationship with the state based on a unique set of entitlements. These entitlements typically involve the inherent right to some form of self-government and an acknowledgment of sovereignty. (Fleras and Elliot 1992:1)

The aim of this chapter is to conduct a comparative analysis of the impact of industrial development on the Khanty, Mansi and Yamalo-Nenets of Northwest Siberia and on the Lubicon Lake Indian Band of Alberta within the context of this process of change from aboriginal peoples to aboriginal nations.

I argue that the degradation of Khanty, Mansi, Yamalo-Nenets and Lubicon territories because of industrial development, particularly oil and gas exploration and extraction, and forestry, has led to the social and cultural deterioration of the Khanty, Mansi and Yamalo-Nenets and the Lubicon Cree. As a result of this siege by governments and big industries, these indigenous peoples are forced to protect whatever unspoiled territory and culture they have left, thus forcing them to redefine their identities as Native peoples and indeed to redefine their relationship with the state. Concomitantly, political awareness and the opportunity to mobilize has led to their politicization as they demand special status as "nations within" (Fleras and Elliot 1992:3).

Historical Background

Khanty, Mansi and Yamalo-Nenets

The Khanty, Mansi and Yamalo-Nenets are three of the thirty-four officially recognized "Small Peoples of the North"—the aboriginal peoples of the Russian North and Siberia who pursue a hunter-gatherer or hunter-herder way of life. The Khanty and Mansi are both Asiatic peoples whose languages belong to the Finno-Ugric language group. They occupy the territory east of the Ural Mountains near the Arctic Circle on the rivers Ob' and Irtysh. First contact with European Russians may be traced back to the eleventh century, when they began trading their furs with the traders from Novgorod (Balzer 1983:636). The tundra-dwelling Yamalo-Nenets speak a Uralic language which is remotely related to the languages of the Finno-Ugric family. While these languages are from the same language group, there are great differences among them from region to region. As is true for the Khanty and Mansi, these differences are so great that the dialect in one region is incomprehensible in another. The Yamalo-Nenets' material culture is very similar to that of other Arctic peoples such as the Yukagir and the Chukchi. These similarities are attributed to the fact that they live in very

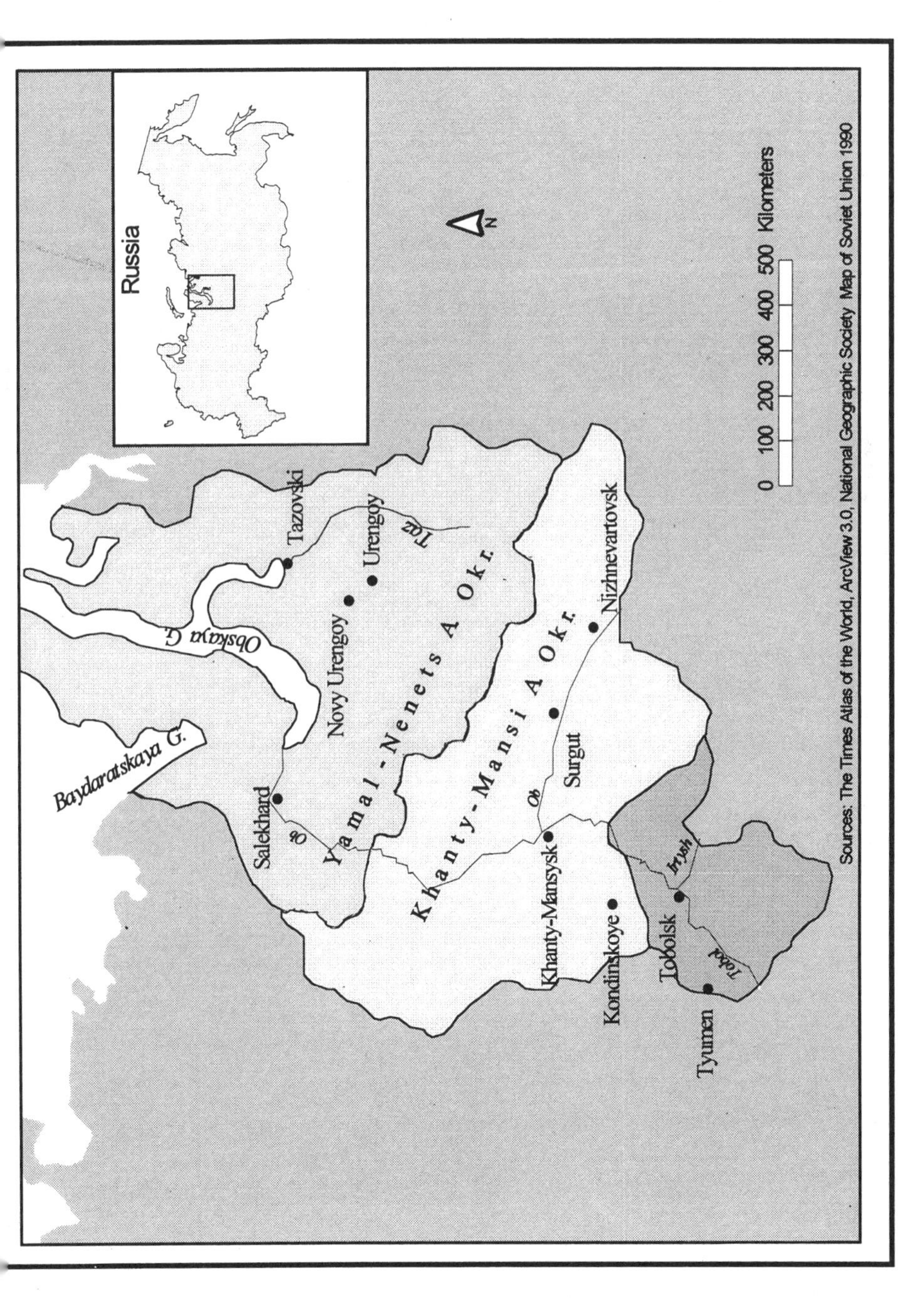

MAP 3.1 Tyumen' Oblast'

similar climatic, environmental and physical conditions. Common among these Arctic peoples is their reliance on reindeer. For the Yamalo-Nenets, both wild and domesticated reindeer have been an integral component of their economy.

As non-European peoples, the Khanty, Mansi and Yamalo-Nenets were seen as inferior races by the Russians, and were therefore exploited for their goods and their resources. Forcible Tsarist jurisdiction over Khanty, Mansi and Yamalo-Nenets territory began in the sixteenth century. Formal bureaucratic jurisdiction over this territory east of the Urals began by building towns, garrison administrations and imposing *yasak,* or payment of tribute in furs (Forsyth 1992:44). All Native peoples of Siberia were forced to pay yasak. Throughout the sixteenth and seventeenth centuries and later, "Russian merchants and rich peasants" forcibly took Khanty, Mansi and Yamalo-Nenets land (Prokof'yeva, et al. 1956:515). Russian peasants turned Khanty and Mansi lands into farmland with the approval of the Tsarist government, while exploiting the territories of reindeer-herding Yamalo-Nenets and northern Khanty for precious furs and fish. Native peoples' fishing and hunting grounds were also taken over by Russians who overfished the rivers by fencing off their mouths to prevent fish from going up river, and overhunted fur-bearing animals for commercial purposes (Prokof'yeva, et al. 1956:515).

Throughout the eighteenth century, the exaction of exceedingly high yasak payments forced the Yamalo-Nenets and the Khanty to abandon their traditional economy of hunting and fishing in order to trap sables, and later foxes, for Russian officials and traders. The Khanty, Mansi and Yamalo-Nenets were, therefore, forced to leave their own territories in an attempt to live as they had lived for hundreds, perhaps even thousands, of years (Prokof'yeva, et al. 1956:515).

These effects on the Khanty, Mansi and Yamalo-Nenets, while serious, were minimal when compared to the imposition of Soviet rule and hegemony. As Debra Schindler argues, Bolshevik policies toward the aboriginal peoples of Siberia soon after the 1917 Russian Revolution revolved around their transformation from "the foraging and herding cultures and economies . . . from archaic forms to socialist forms" (Schindler 1991:70). The Soviet policies toward the Khanty, Mansi and Yamalo-Nenets focused on the goal of transforming their hunting, fishing and reindeer-breeding economy into a socialist economy. To achieve these goals, Schindler contends that the Soviets had to concentrate on four major policies in Siberia:

> the creation of a 'modern,'. . . urban-industrial settlement system; collectivization of the indigenous production economy; development of

> natural resources and the industrial development of other branches of the economy; and the introduction of the indigenous population to and their incorporation in 'modern' (Russian) society. (1991:70)

These aims, based on rigid and dogmatic Leninist ideology, led to the Russification and Sovietization of the Khanty, Mansi and Yamalo-Nenets and other Native peoples of Siberia, and, therefore, to severe damage to their cultures, ethnic identities and ways of life.

The Lubicon Cree

The Lubicon Lake Band of Bush Cree Indians reside in the boreal forests of northern Alberta. Lubicon have inhabited the territory consisting of 7,500 square miles, north of Lesser Slave Lake, southwest of Lake Athabasca, between the Wabasca and Peace rivers for hundreds of years. They traditionally subsisted on moose and other big game and fish (Smith 1988:61).

Initial contact with Europeans dates to the mid-seventeenth century, when the Cree participated in the fur trade (Tobias 1991:214). It was not until this that the Lubicon began hunting and trapping of fur-bearing animals. Until recently, hunting and trapping for fur, plus subsistence hunting and fishing, allowed the Lubicon to support themselves.

Ethnographically, the Lubicon have distinguishing social and anthropological qualities, such as

> bilateral cross-cousin marriage; kinship system with Iroquois-type cousin terminology; classification of kin into consanguine and potential affines or in-laws and including temporary matrilocal (uxorilocal) postmarital residence with bride service; levirate and sorrorate formation of society into hunting groups; local and regional bands; a "marriage universe" including the adjacent Loon Lake and Cadotte Lake regional bands. (Smith 1988:61)

The Lubicon's oral traditions consist of genealogies, historical narratives and Wissakedzak (whiskey jack). Elders play a key role in Lubicon social organization for they are considered wise and resourceful. "Government was by consensus, especially of the heads of families guided by the Elders" (Smith 1988:61).

Like the Khanty, Mansi, and Yamalo-Nenets and other Arctic and subarctic peoples, the Lubicon have an ancient respect for the bear and a "respect relationship for the moose" (Smith 1988:61). Their hunting practices were and still are based on conservation, whereby they will not kill moose calves, pregnant cows or all members of a beaver house. It is kept in mind that these animals must be able to replenish themselves in order to provide food for the Lubicon at a later date. The hunting of moose is

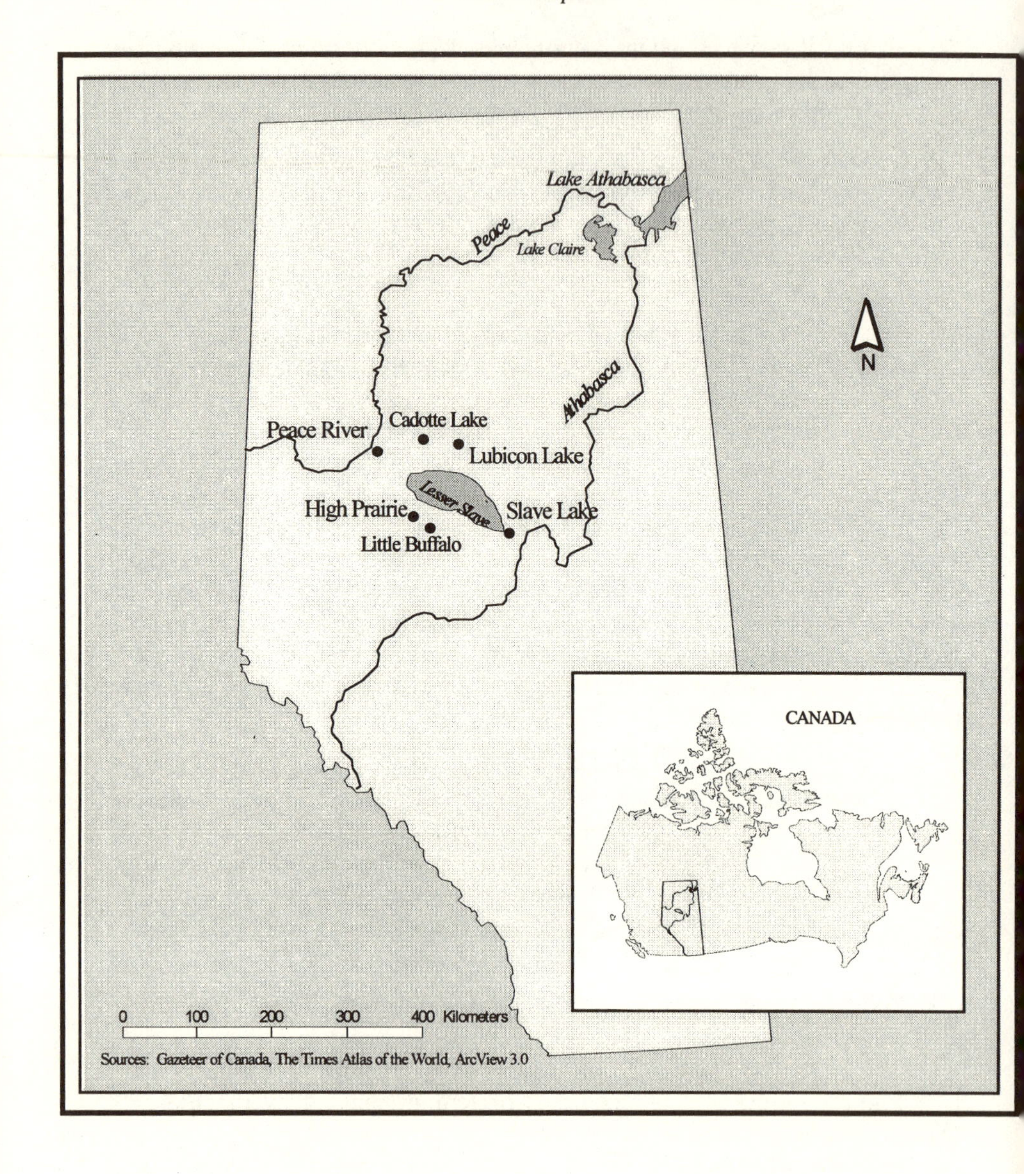

MAP 3.2 Alberta

particularly important because it provides the hunter recognition and respect. The ethic of sharing the kill is also central.

Although the Lubicon Lake Cree participated in the fur trade, European influences were minimal. For the Lubicon this meant only a gradual change in culture and a minor degree of syncretism. They were so isolated that when the Canadian government initiated the signing of treaties in the Athabasca District in 1899-1900, the Lubicon Cree were missed because they did not live near the water routes that the treaty commissioners were surveying (Dickason 1992:390-91). Because the Lubicon did not sign any treaties with the Canadian government, in principle, they maintain control over their territory. While this might seem to provide the Lubicon an advantage, what it means at the present is that they are not recognized to be the rightful inhabitants of the territory they have occupied for hundreds of years. At a presentation to the Standing Committee on Indian Affairs and Northern Development, the Lubicon argued that they had been missed by the Treaty 8 commissioners, and therefore had not been listed under a reservation in the Lubicon Lake area. Instead, because of bureaucratic bungling and obvious disrespect for the Lubicon Cree, both the Canadian and Alberta governments attempted to classify them and, indeed, relocate and enfranchise them, with the Whitefish Lake Band. As Ward Churchill argues, because of oil and gas finds in the 1950s, the Lubicon were "confronted with the specter of complete administrative elimination as an identifiable human group (i.e.: they were, by international legal definition, faced with genocide)" (Churchill 1988/1989:152-174).

Industrial Development: Oil, Gas and Forestry

The polities and political structures that governed Canada and the Soviet Union were very different. Nevertheless, economic and industrial development manifested similar strategies and processes. In the development of resources such as oil, gas and forest products, Canada and the former Soviet Union were similar in their disregard for indigenous peoples and the lands and environment in which they lived. This section illustrates this by reviewing the process of resource development in northwest Siberia and north-central Alberta since the 1950s.

Northwest Siberia

The discovery of oil reserves in northwest Siberia occurred quite by accident in September 1953, when barges transporting rigs were delayed on the Ob' River near the village of Berezov and a test boring by the banks produced a shot of gas and water. Even then, the extent of the oil and gas reserves were not known. Indeed, in a proposal for the future economic and cultural development of the Far North drafted in 1956, the recommendation to extract oil from the Kondinskoe and Surgut regions occupied only three lines of the twenty-five-page document, which made far more detailed recommendations regarding the collectivized traditional economies of hunting, fishing and reindeer-herding activities of the Khanty, Mansi and Yamalo-Nenets (Prilozhenie 1956).

By the early 1960s, the outlook was quite different. It was with great enthusiasm that massive oil and gas development, began in West Siberia, with the implementation of Soviet policies to exploit lands not before utilized for industrial development or resource extraction.[1] More deliberate explorations followed the Berezovskoe gas field find, and by 1960 oil was extracted from Shaim on the River Konda, directly in the middle of traditional Mansi territory. Subsequently, pipelines were built to carry oil to Perm and Tyumen', and railways were constructed to accommodate the migration of new settlers as well as the passage of equipment for oil drilling and natural gas extraction. Specifically in Khanty-Mansiisk Autonomous Okrug, the small villages of Samotlor and Surgut "became centres of intensive oil-drilling operations" (Forsyth 1992:390); and subsequently, in the late 1970s, Yamalo-Nenetskiy Autonomous Okrug, Urengoy and Yamburg accommodated massive gas production. These prodigious industrialization policies were initiated by Nikita Khrushchev to facilitate the Cold War competition between the USSR and the United States, and were enthusiastically continued and intensified by Leonid Brezhnev. Overzealousness regarding output and overestimation of oil and gas reserves in Northwest Siberia would set the tenor of oil and gas development.

For the most part, the expectations regarding the reserves of oil and natural gas in western and northwestern Siberia were fulfilled, especially in the early years of production. However, very much like the history of the oil and gas industry in the Tsarist period, the industry was plagued by inefficient policies that emphasized increasing production at all costs, often neglecting the need for an infrastructure of pipelines to transport the oil out in the first place, not to mention environmental assessment studies. In the 1960s, Northwest Siberia itself lacked the refineries and petrochemical plants to process the oil and gas extracted (Elliot 1974:14). Oil and gas workers themselves were overlooked, and often not provided with adequate housing or social services in the most remote regions of West and Northwest Siberia. Today, the major problem facing the oil and gas industry is finding sufficient investment to modernize outdated drilling and extraction methods. Thousands of kilometers of inferior pipelines that cannot withstand harsh weather conditions require replacements. Amid these infrastructural problems is the immeasurable damage done to the natural environment and the traditional territories of the indigenous peoples who inhabit Northwest

Siberia due to oil spills, gas field fires, abandoned and rusting equipment littering the tundra and taiga, the irreparably scored permafrost, the damaged lichen fields, etc.[2]

Another factor that led to the pollution of the environment and of aboriginal lands in Northwest Siberia was central economic planning itself. In the late 1960s and 1970s, as Brezhnev consolidated his power, the main concern was fully developing the oil and gas industry. There was a great deal of concern for the maximum output from the oil wells, and later the gas wells, but decision-making was not directed by market forces, but instead by bureaucrats in Moscow who may have only known that production meant revenue for the USSR. Quotas set did not reflect the realities of oil production and often showed no foresight. For example, in the 1970s, production quotas on exploration required number of feet dug rather than how much oil was actually found. "In this situation the crew is likely to overdrill a discovery in order to meet its footage assignment" (Campbell 1968:84). This activity scarred the taiga and tundra and littered the landscape with equipment from the test drilling, whether or not the various sites yielded any oil or gas.

The drive to discover more sources of oil and gas in Northwest Siberia was also an attempt to meet the Soviet population's and industries' growing demand for energy, to offset the declining oil production in other parts of the Soviet Union and, especially in the late 1970s, throughout the 1980s to the present, to provide much-needed hard-currency to the government in Russia. By the early 1970s, as oil and gas reserves were being depleted in European Russia, the demand for other sources of oil and gas lay heavily on West Siberia. As a result, by the early 1990s, West Siberia was producing well over sixty percent of the oil and gas needs of the Soviet Union (Bradshaw 1992:13-15).

While the complaints of oil and gas workers were heeded, most notably by the Gorbachev government, the voices of indigenous peoples were not. Arguably, the Soviets realized that oil meant power and thus was central to the maintenance of the Soviet Union as a superpower[3] and, today, to the development of the Russian Federation as a viable and strong democratic nation-state. As Robert E. Ebel avers,

> Oil is a high-profile commodity, fueling much more than automobiles and airplanes. Oil fuels military power, national treasuries, and international politics. Because of this it is no longer a commodity to be bought and sold within the confines of traditional supply and demand balances. Rather, it has been transformed into a determinant of well-being, of national security and international power for those who possess it and the converse for those who do not. (1994:14)

For the Soviet regime under Brezhnev and now for the Russian government, the major source of oil and gas would be in Tyumen' Oblast', West Siberia. It would not be an exaggeration to assert that the regimes of both Mikhail Gorbachev and Boris Yeltsin operated and operate within this framework. What this means for the Khanty, Mansi and Yamalo-Nenets of Northwest Siberia is that the specter of great-power politics has irreversibly changed their lives.

Gorbachev attempted with limited success to reform the oil and gas industry by overturning some of the policies instituted by Brezhnev, whom Gorbachev blamed for the crisis situation in the energy sector of the economy, and indeed the economy as a whole. The major mistakes made in the energy sector, as Gorbachev saw it in the late 1980s, were that the enormous investment in the natural resources was haphazard and "inordinate," with the hard currency profits used to patch up the economy instead of for long-term investments in the modernization of the economy (Gustafson 1989:6). While certain measures were put in place to resolve some of these problems, including addressing the needs of oil and gas workers, and decentralizing the industries through *khozraschet* (cost-accounting), for the most part the strategy of revitalizing the energy sector was the same. "Gorbachev's approach to energy policy is so far almost indistinguishable from Brezhnev's in its two most dangerous aspects: the rapid rise of energy investment and the failure to curb energy consumption" (Gustafson 1989:7). In the end, however, it was evident that while the concerns of indigenous peoples were paid lip-service, the reality was that the reconstruction of the Soviet Union into a modern nation-state proved to be too deeply rooted in the heavy exploitation of fossil fuels in West and Northwest Siberia.

Despite the declining output in the late 1980s and early 1990s, West Siberia remains the main source of fossil fuel energy in the Soviet Union and after 1991, the Russian Republic. Mikhail Gorbachev understood this well.[4] Indeed, the potential is there, with fully one-third of all gas reserves in the former Soviet Union lying beneath the Yamal peninsula, territory designated by reindeer-herding Khanty and Yamalo-Nenets to be prime grazing land for their reindeer (Osherenko 1995:225-237). The aboriginal minorities who regard major tracts of Northwest Siberia as their traditional territory have very limited options in a Russia reliant on West Siberia and the Yamal not only for energy but also for hard currency and trade commodities (Nulty 1992:126-130).

For the indigenous peoples of Northwest Siberia, the development of oil and gas reserves on their traditional hunting, fishing and reindeer-herding territory transformed and re-identified it as empty space to be populated, conquered and exploited at whim, as a place to build massive cement cities and to drill hundreds of thousands of possible oil and gas sites, and as a major source of hard-currency revenue. And they, too, would be transformed and recreated into the industrialized *Homo Sovieticus* ("L" 1959; Zubek 1984; Tasch 1996), altering their aboriginal identity possibly beyond restoration.

Northern Alberta

In 1914, oil was first discovered in Alberta in the Turner Valley, west of Calgary. Output from this area was sporadic, characterized by thirty years of boom and bust. What truly opened up the oil industry in Western Canada, particularly Alberta, was the discovery of oil in Leduc just south of Edmonton in 1947, followed by remarkable discoveries of oil and gas in British Columbia, Saskatchewan and other parts of Alberta. These oil finds resulted in major investment by the world's largest oil companies with only twenty percent of the investments coming from within Canada (Laxer 1983:6). "Canada's petroleum industry had become a northern adjunct to the American industry" (Laxer 1983:7).

Development of the oil and gas industry in the Lubicon territory in northern Alberta began in the 1950s and, like Northwest Siberia, was comparatively minimal with only eleven wells drilled. However, this number doubled in the 1960s and doubled again in the 1970s. In 1979, with the oil crisis precipitated by the Iranian revolution, the oil and gas on Lubicon territory became more valuable, and even more development ensued with wells built east, south, and northeast of Little Buffalo.

Pierre Trudeau's administration brought in the New Economic Policy in 1980, making the oil on Lubicon lands even more precious and a major potential for the province of Alberta's and Canada's revenues. The New Economic Policy allowed oil prices to rise every six months to catch up to world prices. It also established a two-tiered pricing system that "encouraged new development: one price on oil discovered before 1981, a higher price on new discoveries" (Goddard 1991:76). This pricing policy spurred new activities of exploration and drilling in Alberta, especially in the territory claimed by the Lubicon Cree. Because most of the new discoveries of oil on Lubicon territory came after 1981, oil companies and the government benefited from the favorable prices. As Pratt and Urquhart (1994:111) contend, "the province did nothing to stem the occupation of Lubicon territory by the energy industry; in fact, government encouraged it."

Like the Russian government's treatment of the Natives of Siberia, the governments of Canada and Alberta have chosen to ignore the legitimate concerns of the Lubicon Cree, who have witnessed the construction of a road through their territory, the exploration and extraction of oil and gas by major oil companies, as well as the harvest of prime lumber from their lands by the Japan's second-largest pulp and paper company, Daishowa. Under the Alberta government's aims of economic diversification and exploitation of its forestry resources, Daishowa answered a call to develop an "Alberta-based pulp and paper industry" (Tollefson 1996:122; Pratt and Urquhart 1994:43). Now not only was Lubicon traditional territory under siege from oil and gas development, but in 1988 Daishowa was given the right to build a pulp mill to be located sixty-five kilometers south-west of Little Buffalo. This multinational company was also given a license to manage some 29,000 square kilometers of forests as a source of timber for the mill which now has a processing capacity of 11,000 trees per day.

Daishowa's authority to manage the forest for its own pulp mill was given without consulting the Lubicon Cree Indian Band, whose territory is within this Daishowa-managed forest (Tollefson 1996:122). Opposition to the deal between the Alberta government and Daishowa was swift. International and national support went out to the Lubicon and to Chief Ominayak, pressuring Daishowa to meet with them in March 1988. The results of the meeting are contested. The Lubicon understanding is that Daishowa would stop logging in Lubicon territory and that Lubicon concerns about forest management and wildlife would be heeded until the Band's land claim was settled. Daishowa denies that such an agreement was ever struck with the Lubicon (Tollefson 1996:123). In the absence of any agreement or understanding between the Lubicon and Daishowa, Daishowa started clear-cutting timber on contested Lubicon land in the fall of 1990 with the intent of continuing through the winter of 1990-91. This served to worsen the already tense relationship between Daishowa and the Lubicon.

Problems of Social and Cultural Decline

Northwest Siberia

Major industrial development led to the significant increase in non-indigenous population in the Khanty-Mansiisk Autonomous Region from 98,000 in 1959 to over 500,000 in 1979, and by 1989 to almost 1 million (Gosudarstvennyi Komitet SSSR po Statistike 1991:46-48). By contrast the Khanty and Mansi population increased only by twelve percent to 28,497 in the same period. In 1959, the Khanty-Mansiisk region was seventeen percent Khanty and Mansi, by 1979 their share of the population in their own region fell to less than four percent (Pika and Prokhorov 1989:76-83). Today the Khanty and Mansi number only one percent of the population on their own traditional lands (*Current Digest of the Soviet Press* 1990:20-21). In the Yamolo-Nenetskiy Autonomous Region, indigenous peoples make up six percent of the population. Vladimir Sangi, the first President of the

Association of the Small Peoples of the North, stated in somewhat romantic tones,

> The other civilization burst into this fragile civilization with all its energy and, like a tank, rumbled over the body of the northern culture that was incomprehensible and alien to it. And as long as we keep going on about how many schools and hospitals have been built in the North during the past 70 years, which peoples have become literate, how many people have become doctors, teachers, writers, etc., without taking into account what has actually happened to entire peoples, we won't find a way out of our historical impasse. (*Current Digest of the Soviet Press* 1990:20-21)

The urgency of the situation of aboriginal peoples in Siberia is certainly evident in Sangi's statement, underlying the immediacy with which action must be taken to save the lost language, culture and traditions of the Native peoples of Siberia, including the Khanty, Mansi and Yamalo-Nenets.

Those who came to Northwest Siberia were there for fast money and their sole purpose was the exploitation of the land and its resources. They had very little concern for the natural environment that was Khanty territory. Migration into western Siberia by Europeans has led to problems of a highly transient population committed only to making the highest wage they can find (Bond, et al. 1991:363-432). Moreover, speaking of Siberia as a whole, "an additional factor contributing to labor movements within the region are boom-and-bust economic cycles, to which Siberia is vulnerable because of the unidimensional, raw-material orientation of its economy. Mineral deposits and timber stands are exploited and abandoned as production shifts to new locations" (Bond, et al. 1991:363-432).

While the European population came to the Tyumen' Oblast' by the tens of thousands, the Khanty, Mansi and Yamalo-Nenets left from the growing urban cores of the region. Echoes of this date back to the beginning of Russian colonization of Khanty, Mansi and Yamalo-Nenets lands. In an attempt to preserve their traditional way of life and customs, the Khanty, Mansi and Yamalo-Nenets migrated away from centers of Russian industry and trade, continuing and maintaining their life of hunting, fishing and reindeer-herding.

While some Natives hold non-traditional occupations in the fields of medicine, education, and in more recent years, politics, for many the Soviet lifestyle is antithetical to their social and economic organization. Many, as they did in the seventeenth century, are still migrating away from the urban industrial centres that accommodate the oil and gas industry in the Khanty-Mansiisk and Yamalo-Nenetskiy Autonomous regions. The result is that the Khanty, Mansi and Yamalo-Nenets live in substandard existence, not being able to do what makes them Khanty, Mansi and Yamalo-Nenets while they see their territory pillaged by both Russian and Western interests. Presently, "in the economic balance of the region, the production generated by the indigenous northerners,. . . has become almost unnoticeable against the huge industrial capacity. The situation in the Khanty-Mansi autonomous region is quite typical for all the North" (Pika and Prokhorov 1989:124).

The modernization and industrialization of the Natives of Northwest Siberia and their traditional territories have not led to their progress as a people, even if it has made them citizens of the Soviet state. Rather it has led to culture loss, hardship and decline in their standard of living. Moreover, this industrial exploitation of Khanty, Mansi and Yamalo-Nenets land has concurrently led to the exploitation of those who decide to stay and work in the urban centres, or even those who may just live around sources of oil and gas. In the late 1980s, "Ethnographers [were] alarmed by the crisis they discern[ed] among Native Siberians, whose living conditions are acknowledged to be the worst in the USSR" (Mihalisko 1989:3). An interview with an executive of the Salekhard city government revealed that Yamalo-Nenets within sight of the city live in squalid conditions. Most of these Yamalo-Nenets prefer to live in *chums* (triangular tent-like dwellings traditionally made of reindeer skins) and to practice their traditional economies. But for most, owning reindeer was and is not possible, and hunting and fishing is not enough to sustain them as rivers and lakes become more polluted, and hunting and gathering grounds are overtaken by oil and gas development. Therefore, many Yamalo-Nenets living near Salekhard are forced to ask for government assistance despite the abundance of oil and gas on their territories.[5]

Soviet population specialists Aleksandr Pika and Boris Prokhorov assert that in Siberia less than forty-three percent of "the working-age population . . . is engaged in the traditional occupations of hunting, fishing, and reindeer husbandry, as compared to seventy percent thirty years ago" (Mihalisko 1989:4). The decline of the Native traditional economies is staggering, especially when it is evident that these traditional forms of labor are being replaced by non-traditional ones. Those who choose to keep the traditional forms of labor are not allowed to freely hunt or to herd reindeer as they had done traditionally. Rather, they are assigned into residence areas called "national settlements" (Mihalisko 1989:4), and as in the past people are usually placed in the strict collectivized settings of the *sovkhoz* (state farms). Soviet policies regarding the industrialization of the north and its population would force a new identity upon the Natives of Northwest Siberia. It would force them to become the industrialized Soviets the Marxist-Leninist doctrines prescribed, no matter how unsuccessful the outcome. Towards the end of the 1980s, as oil and gas development went into further decline, and as the promises of a socialist utopia grew even further from reality than when the experiment began, industrialization could no longer provide for the Khanty, Mansi and Yamalo-Nenets. And today, still grappling with the Soviet legacy, the sovkhozy themselves are no longer able to provide many Natives in Khanty-Mansiisk and Yamalo-Nenetskiy Autonomous Okrugs with income to allow them to sustain a decent standard of living.

While oil and gas production in West Siberia has been fluctuating since the early 1980s,[6] the impact of oil and gas exploration and extraction has meant a steady decline in the standard of living, health, and retention of language, culture and traditional economies of Native peoples. Exacerbating this decline is the degradation of the environment on which these indigenous peoples depend for their economic livelihood, their food, their cultural and spiritual activities, and the maintenance of their traditions.

Northern Alberta

The Lubicon, too, have suffered greatly due to the disrespect with which they are regarded by the provincial and federal governments. This disrespect has culminated in trespasses on Lubicon land in order to establish an infrastructure so that industrial development in the territory may be initiated. Failing in their attempts to designate their territory as a reservation, the Lubicon saw their land unceremoniously divided into "leases for exploration and exploitation to major oil companies" (Smith 1988:62). Indeed, by 1984 more than 400 oil wells had been deployed within a twenty-four kilometer radius of Little Buffalo, the center of the Lubicon band community (Dickason 1992:391). Moreover, to exacerbate the problem, the all-weather road going through traditional Lubicon Cree territory was completed in 1978 without any study of the environmental impact. This marked a significant turning point in the continued atrocities against the Lubicon.

Seismic lines and roads that were subsequently built bulldozed over the registered traplines of the Lubicon, "destroying traps and the potential of a regular harvest of fur from forest animals" (Smith 1988:62). The activity of the oil companies in the Lubicon Lake territory has driven out the moose, the prime source of game for the Lubicon. In addition, companies and the Alberta government marked the new roads criss-crossing through the Lubicon lands "Private Road" and thus the Lubicon were not permitted to hunt or trap on their own traditional territory.

This lack of access to the land resulted in the destruction of their subsistence economy. Predictably, these developments had terrible consequences, leading the Lubicon for the first time to seek welfare benefits from the Canadian government so that they could survive. Indeed, welfare rates increased from ten percent to ninety percent in the period from 1979 to

1984 (Pratt and Urquhart 1994:111). Ironically, as Chief Ominayak notes, "with the billions of dollars that have been extracted by way of natural resources off our traditional territory, there has not been a red cent that has been coming back to the community other than welfare from the federal government" (*Lubicon Settlement Commission* 1993:13). The Lubicon lost their autonomy and control over their own lives, as well as the dignity of being able to sustain themselves through their traditional economies of hunting, fishing and trapping. No amount of social welfare provided by the Canadian government could replace what the Lubicon lost. This, in turn, led to the loss of human dignity, exacerbated by degradation of the traditional environment from which the propagation of culture begins.

The industrialization of the traditional Lubicon territory without the consent of the Lubicon left them ill-prepared for the social and cultural problems which ensued, such as alcoholism, violence, and other social pathologies. Once a self-sustaining community, relying on hunting as its major economy, it now depends on government handouts and welfare. In 1992, Crystal Gladue, a 14-year old girl from Little Buffalo remarked: "In the last few years, since we were young, we have seen more troubles here. More alcohol and with it fights and accidents. People don't get along any more [sic] as well as they used to. People from outside come and sell booze and it breaks up families and causes violence" (*Lubicon Settlement Commission* 1993:16). The influx of more people into Lubicon territory has also increased their exposure to diseases such as tuberculosis. In the August 1987 alone, forty-five cases of tuberculosis were found among the Lubicon, leading to the hospitalization of thirty-two of them (Smith 1988:62). Because of this tuberculosis epidemic, Friends of the Lubicon was formed by a group of concerned individuals who visited Little Buffalo in 1988. The aim of the group was to help the Lubicon out of the bad conditions in which they lived, to help them overcome tuberculosis, and to help the Lubicon settle land claims (Tollefson 1996:123).

The devastation caused by the oil, gas and forestry infrastructure is also illustrated in the testimony of the witnesses for The Lubicon Settlement Commission of Review. As Violet Ominayak testified, "we live in constant danger. . . . The roads are dusty and dangerous to travel. The logging and oil trucks run us off sometimes. We have lost many young ones because of the horrible roads." Continuing with her testimony, Violet Ominayak strongly voiced her demands: "The Lubicon women demand an end to the physical, emotional, economic, cultural and spiritual destruction. Hear our voice and our message—we don't know if we'll be here tomorrow" (*Lubicon Settlement Commission* 1993:11). Despair, frustration at the situation, and demands for justice characterized many of the testimonies at the Review hearings. Behind the colonized voices silenced for so long are Lubicon individuals urgent to be heard and taken seriously. By expressing these demands, the Lubicon become more and more politically aware, clamoring for land and for recognition amid the fear of losing their traditional territory and their people.

From "Aboriginal Peoples" to "Aboriginal Nations"

For the Natives of Northwest Siberia and northern Alberta, there is much work to be done before their rights as indigenous societies are recognized and secure. Nevertheless, they are successfully drawing attention to their plight. Through lobbying and protest they have begun to transform "relationships of domination and control" into "patterns of interaction that claim to embrace the unique status of aboriginal peoples as 'nations within': that is, [as] distinct societies with special claims and collective entitlements that derive from formal recognition as the indigenous occupants of land" (Fleras and Elliot 1992:3). Caution must be exercised, and it must be stressed that the Khanty, Mansi, Yamalo-Nenets and Lubicon are just beginning their journey towards designation as 'nations within.' But the processes of change from domination to interaction that Fleras and Elliot write about are evident. The Khanty, Mansi, Yamalo-Nenets and Lubicon are no longer complacent about how they are treated within the dynamics of industrial development pursued by Russia and Canada. In this refusal to be complacent are the seeds of political action and mobilization so necessary for the transformation from aboriginal peoples to aboriginal nations.

In search of an ethnic identity

What has been happening among the Khanty, Mansi, and Yamalo-Nenets as well as the Lubicons is that with the pillaging and exploitation of their lands by both of their respective governments and by the industrial companies who want to profit from the resources on their territories, a strong desire has arisen among them to return to their traditional cultures, language and way of life (Pika and Prokhorov 1994). Having experienced paternalistic rule by the state and the destruction brought on by industrial development, indigenous leaders in Northwest Siberia and northern Alberta realize that what is important for them is to protect their relationship to nature and to the land (Goddard 1991; Petrunenko 1991:40-41). The irony, however, is that it is not only land that has been industrialized; the Khanty, Mansi, Yamalo-Nenets and Lubicon Cree themselves have been industrialized. This battle to recapture and maintain their traditional territories and cultures has been carried out in the political arena, politicizing the relationship between the dominant polities and the Native peoples of northern Alberta and Northwest Siberia. "Aboriginal national concerns are varied, but most tend toward the principle of self-determination at political, social, and cultural levels as part of a wider drive to redefine their relationship with the state" (Fleras and Elliot 1992:1).

For the Khanty, Mansi, and Yamalo-Nenets the primary goal is the maintenance of their ethnicity and their culture. They realize that their survival as a people can only be insured through the propagation and promotion of their way of being as superior, or at least equal, to the hegemonic polities and societies by which they are subjugated. This heightened sense of their identity as indigenous peoples with distinct ethnicities is the weapon against the annihilation that they fear.

The Khanty and Mansi seek the protection of reservations in order to preserve their way of life. The Khanty and Mansi believe that the designation of reservations, although it does have a pejorative connotation (especially in the West), would lead to a preservation of their territory and the Khanty and Mansi ethos. Having seen the destructive nature of industrial development and the apparent road to extinction which it represents, the Khanty and Mansi postulate a solution that suggests a heightened sense of their indigenous identity and the opportunity to pursue their way of life within the social and political organization of Khanty and Mansi culture and ethnicity. Oksana Petrunenko argues that the idea of establishing a reservation may improve their situation, and indeed scholars, philosophers and lawyers involved in Northwest Siberia support "the idea of setting up an ethnic territory or a reservation," so long as it corresponds with the "international agreements on the rights of peoples leading a traditional way of life" (Petrunenko 1991:40). The reservation they seek would protect their land and their people from the negative impact of industrialization of the region, because "priority is to be given to the traditional crafts, hunting and fishing. As on any reservation, the number of outsiders, such as tourists, builders and oil workers, will be sharply limited. The prospecting parties working there will move elsewhere" (Petrunenko 1991:40). Indeed, one-third of the Khanty-Mansiisk territory was declared a reservation, stopping any logging or industrial development within this 100,000 square-kilometer reserve. However, much as was the case in the Soviet period, the mandated reserve exists only on paper.

In northern Alberta, again as a consequence of being overlooked by the Treaty 8 commissioners, the Lubicon are not given jurisdiction or title to their lands, meaning that they cannot make any land-use decisions nor do they have any input into the development of their own territory. Fearing the encroachment of the dominant culture, the Lubicon have been requesting a reserve since 1940. The response by the Alberta and Canadian governments has been to stall. Since the size of reservations is determined by the number of members in the band, "there has been ongoing controversy over the number of registered members in the Lubicon Lake Band" (Cloutier 1988:1-17). The situation is not made better by the fact that Lubicon Cree territory is rich in oil, gas and timber.

Despite the more than fifty-year battle fought by the Lubicon, far from giving up on their demands, since 1978 their resolve to obtain what is rightfully theirs has strengthened. This fight for a reservation and for control of resource rights over Lubicon territory is personified in the Band's chief, Bernard Ominayak. "Chief Ominayak feels the Alberta government is trying to wear down their spirit until they give up their will to survive on land they have a right to have" (*Sweetgrass* 1984:13).

Also personified by Chief Ominayak is the entire Lubicon Lake Indian Band's newly voiced ethnic self-consciousness. Over the last fifteen years, Chief Ominayak has steadily advanced the cause of his people. Often described as unassuming and quiet, (Goddard 1991:1) Ominayak took on the role of Chief just at the height of the oil boom in Alberta, just as the all-weather road was completed and just as seismic and drilling crews seemed to be overtaking Lubicon lands. Ominayak took on the challenge, however, and through his battle for his people learned about the politics of land claims and land rights for aboriginal peoples, preparing him for even greater battles with provincial and federal levels of government.

Politicization

It may be argued that the modern Khanty, Mansi and Yamalo-Nenet culture and ethnicity no longer exist in their pre-contact form. However, the attempts to imagine, recreate and reinvent traditional customs, ways of life and environment is strong among indigenous elites. With the policies of *glasnost* (opening) and *perestroika* (restructuring) introduced by Gorbachev in the Soviet Union in 1985, and then the Union's subsequent collapse, Khanty, Mansi and Yamalo-Nenets elites have become politically empowered to counter their destruction at the hands of the Soviet hegemonic power, and now, at the hands of economic and environmental mismanagement of the oil and gas industry in the Khanty-Mansiysk and Yamalo-Nenetskiy Autonomous regions. Using this opportunity to voice their political concerns, these indigenous elites have mobilized the resources at their disposal and within their reach in order to draw attention to their demands and to place them on the political agenda of the Russian government.

How are the Natives of West Siberia expressing this new-found politicization? The creation of such organizations as the Association of the Small Peoples of the NorthAssociation of in 1989, now headed by the Khant writer Yeremei Aipin, is but one form of this expression. The Association advocates the protection of Native rights and resources in accordance with the United Nations' mandates regarding indigenous peoples, including an effort to preserve the environment to which their traditional economies and cultures are intrinsically tied. The association seeks to promote Native peoples' governance of themselves and their territories (IWGIA 1990).

Local-level associations form another means by which Natives have joined the political arena. The Khanty and Mansi formed *Spasenie Yugri,* and the Yamalo-Nenets *Yamal-Patomkam.* Both organizations assert political mandates that strongly advance the return to traditional culture, language, economies for Native peoples. Indeed, while each association has a different way in which to achieve their aims, both see that the way to preserve and develop their peoples is to assert themselves legally as indigenous peoples, with special land and resource rights under international law.[7]

It may be argued, therefore, that the Khanty, Mansi and Yamalo-Nenets are reformulating their identity from Soviet citizen to indigenous peoples with distinct rights, indeed as global citizens. In the case of *Yamal-Patomkam*, the problem, as Norman A. Chance and Elena N. Andreeva assert, is that with scant political experience and "no recognized legal base, [their] political power is severely limited" (1995:217-240). With *Spasenie Yugri*, it is evident that there has been no coordination between the head office in Khanty-Mansiisk city and the smaller villages. This has produced a great deal of resentment and skepticism that the association would do anything for the local Natives. Many intimated that upper-level elites such as Yeremei Aipin were too far removed from the people.[8]

The election of the indigenous Khanty leader Yeremei Aipin[9] to the Peoples' Deputies of the Russian parliament is a signal that Native peoples are taking their fate into their own hands, and indeed want a representative in the Russian Parliament and are willing to address Moscow within the Russian Parliament (Aipin 1991). This is a strong sign of politicization that may lead to the preservation not only of the environment, but also of Native culture and ethnicity.

Besides the barricades and sometimes violence, the petitions to governmental and non-governmental agencies, and the attempts to educate the public to the concerns of these respective peoples, the Khanty, Mansi and Yamalo-Nenets have worked to gain support from governmental and non-governmental agencies. While recognition as indigenous peoples is a step towards putting their concerns on the political agenda of Moscow, ultimately it is the political will of the central government that will give power to the voices of these Native peoples. The reality is that within the framework of a democratizing capitalist government that relies on West Siberian oil and gas for most of its energy needs as well as its foreign currency revenue, the needs of the Khanty, Mansi, Yamalo-Nenets and the environment in which they live is a low priority within the Yeltsin government. Nevertheless, like many marginalized groups who organize, mobilize and protest, the Natives of Northwest Siberia have become increasingly politicized.

The same may be said of the Lubicon Cree who have been fighting to settle their land claims for over fifty years. Also important in the search for ethnic identity in an industrialized world is the concomitant politicization that has developed from lobbying government and industries. The numerous commissions, press conferences and appeals to the United Nations attest to the determination of the Lubicon Cree to solve their landless status through political means.

The Lubicon, in particular, have made their name known throughout the world with the help of Fred Lennarson, an long-time activist for civil rights in the United States and Canada. Since 1980, the Lubicon have initiated legal action against the Canadian Government. Because the Canadian government has been reluctant to give the Lubicon reserve lands for the 450 members that the Band claims it has, they are still fighting for a fair share. The government of Canada insist that only 200 persons are registered as Lubicon Cree. Therefore the struggle continues as the Lubicon loudly protested the Olympic Games in Calgary in 1988, as well as the exhibition on aboriginal peoples at the Glenbow Museum in Calgary because it was sponsored by Shell Oil, one of the companies exploring in Lubicon lands.

In an attempt to impose further pressures on the governments of Canada and Alberta, the Lubicon have gone to the United Nations Committee on Human Rights to have their case heard. In 1988, the UN declared that the Lubicon Lake Band was justified in their claims of abuse of human rights by the Canadian government. Canada was thus ordered "under Provisional Rule 86 to do no further damage to traditional Lubicon society" (Smith 1988:62).

Following the hearing with the UN, not much was done to fulfill the ruling. It did compel the Premier of Alberta at the time, Don Getty, to concede to the Lubicon one-hundred-and-twenty-six square kilometers of land with full subsurface rights and another twenty-five-and-a-half square kilometers with surface rights only. This was a short-lived victory for the Lubicon as the Daishowa Corporation, not seeing fit to place a moratorium on logging of Lubicon land until claims are settled, began clear-cutting just two years later. The Lubicon, in turn, protested by setting logging equipment on fire and by continually appealing to those who may be able to help, whether it is the UN or the Friends of the Lubicon.

Indeed, Friends of the Lubicon have initiated protest against Daishowa and its logging practices. On behalf of the Lubicon, this non-governmental organization asked for a boycott of Daishowa paper products. The aims of the protest were to have Daishowa admit that it had reached an agreement with the Lubicon to stop logging on its contested territory, and to respect the environmental and wildlife concerns of the Lubicon. Friends of the Lubicon solicited support from various groups including the Ontario Liquor Control Board, fast food restaurants, retail and grocery stores, appealing to them to stop using Daishowa paper products (Toffelson 1996:123).

The pressure brought on by the protest and the boycott led Daishowa to reverse its plan to persist in logging Lubicon territory in 1991-92. The Friends of the Lubicon continued with their protest, getting the support of twenty-six companies by the summer of 1993 and almost fifty companies by the fall. These companies represented between 2,700 and 4,300 retail stores (Pratt and Urquhart 1994:18, 126; Tollefson 1996:123). With the pressure of the boycott and protests becoming unbearable, Daishowa launched a lawsuit against Friends of the Lubicon and three of its leaders. Reluctant to concede any more to the Lubicon and its supporters, Daishowa struck back alleging damage due to financial loss because of the protest. In an indirect way, Daishowa's lawsuit is an attack on the Lubicon themselves as they attempt to define their relationship with industries and governments by attempting to settle their land claims. In sum, problems resulting from the dominant power's policies aimed at facilitating maximum profit from the exploitation of resources, not surprisingly, have led to heated conflict between the Lubicon and its supporters, and governments and multinational corporations.

In 1992, negotiations between the Canadian government and the Lubicon began again, but without much results. In June, the Lubicon Settlement Commission of Review initiated its hearings, producing a final report and recommendations in March 1993. The recommendations of the final report clearly favored the Lubicon as it suggested that such agreements as the Grimshaw Accord be ratified, that the extinguishing of aboriginal and land rights not be part of any settlement between the Lubicon and the federal government, and that the compensation asked for by the Lubicon be given (*Lubicon Settlement Commission* 1993:5-6). While this is another victory for the Lubicon and aboriginal rights in Canada, the recommendations have yet to be implemented.

Conclusion

As the Khanty, Mansi, Yamalo-Nenets and Lubicon Cree transform themselves from "aboriginal peoples" to "aboriginal nations," it is evident that they have much to overcome. Colonization and assimilation have taught them to forget their traditions, their languages, their cultures and their ways of life. In exchange, the Khanty, Mansi and Yamalo-Nenets were taught how to be Russian, how to speak Russian, and how to think Soviet. With the collapse of the Soviet system and with the environmental damage done to their territories because of oil and gas development, these Natives are hard-pressed to find ways to again practice their traditional cultures, to speak their Native tongues and to pursue their traditional economies. For the Lubicon Cree, some of the effects of colonization and industrial development are the same. Loss of culture, loss of population due to disease and loss of territory because of resource development plague the Lubicon. These Natives of Northwest Siberia and northern Alberta have only just begun to transform the relationships of domination that they have experienced over the centuries. Nevertheless, there have been victories along the way as their claims and interests are heard over the din of industrialization.

Still, the most formidable task facing the Native peoples of Northwest Siberia and northern Alberta is to define what it means to be Native in a highly industrialized environment, already pillaged and degraded to the point of exhaustion. The task is also to define what the indigenous peoples' relationship is and will be to the central governments in Moscow and Ottawa respectively, to the various local governments and administrations, and to industrialists, so that their interests may be voiced and acknowledged alongside the enormous political and economic forces favoring the exploitation of oil, gas and timber resources on their traditional territories. Defining themselves as "nations within" gives them a political voice within the respective societies, polities and economies in which they now live.

In this transitional economic and political period, pressures for industrial development in the North will increase, by both domestic and foreign developers. Vesting rights to substantial areas of the North in the aboriginal population will likely improve the potential for the development of both traditional activities, and the possibilities for more rationale development of non-traditional activities. Until legislation provides for such comprehensive land claims, the chances for both promoting cultural survival and reducing environmental threats remain imperiled.

---

[1] Reports from geologists and geophysicists to the Geological Department of the Tyumen' Oblast' government boast of drill masters drilling twenty to twenty-five percent deeper than the yearly plan, echoing the height of Stakhanovism in the 1930s. (See, for example, Postanovlenie: preziduma TsK profsoyuz rabochikh geologorazvedochnykh rabot, "Protokol No. 25: Ob initsiative kollektivov burovyk brigad Tyumenskogo upravleniya t.t. Urusova S.N. i Tarasova, A. F." Moskva, 19 marta 1963g.)

[2] Interviews with Yamalo-Nenets in Tibey Saley on the banks of the Taz River in Tazovskii Raion in the summer of 1994 revealed that gas field fires were seen regularly from their village. The Yamalo-Nenets here, especially the men, also complained that they no longer had reindeer because of collectivization policies. The once nomadic

reindeer-herding Yamalo-Nenets who now live in Tibey Saley are forced to live sedentary lives fishing, hunting and gathering berries as major sources of economic subsistence.

[3] Witness the panic that the Brezhnev leadership exhibited when informed in the mid-1970s that not only were energy sources declining in European Russia, but new possible sources of oil in Tyumen' Oblast' had not reached projected estimates. Moreover, there were reports from geologists that oil sources would decline rapidly and steadily by the next decade. Brezhnev's concern was not only to keep industrial output, especially oil production, in line with the Tenth Five-Year Plan, but also with hard currency revenues brought in by selling oil to Western Europe. The investment and policy change in favor of developing the oil and gas industry in Tyumen' Oblast' was immediate (see Gustafson 1989).

[4] Gorbachev also asserted that the successful revitalization of the oil and gas industry in West Siberia would mean the advancement of socialism (Gorbachev 1995:6).

[5] Interview with Tamara Georgiyevna Ananyeva, Director for Social Problems, Salekhard City Administration, Salekhard, Tyumen' Oblast, conducted April 1995, Prince George, B.C., Canada.

[6] In 1983, oil production in West Siberia was 616 mn tons, in 1985 it fell to 595.3 mn tons, but by 1988, spurred on by perestroika and calls for increased production, it rose to 624.3 mn, tons. By the following year had decreased to 606.3 mn tons and in 1990 to 569.7 mn tons (Bradshaw 1992: 16).

[7] By putting together political pamphlets representing their mandate and demands from the Russian government, the Khanty, Mansi and Yamalo-Nenets engage in political debate and political action unheard of under Soviet rule. ("Programma Assotsiatsii 'Spasenie Yugri,'" *Leninskaya Pravda* 20 July 1989: p. 6; and "Yamal-Potomkam," *Ustav: Assotsiatsii korennykh narodov Severa Yamalo-Nenetskii avtonomnogo okruga (respubliki) "Yamal-Potomkam!"* Salekhard, 31 January 1991.)

[8] Interview conducted with Mansi Natives in Listvyenichnyi in the summer of 1993. These sentiments were echoed by two lower-lever Khant elites, Vladimir Kogontchin, Head of National Khant Obshchina in Ugut and Petr Moldanov, Chairman of National Khant Obshchina in the Surgut region. Interviews and conversations conducted in Prince George, B.C., Canada between 25-29 May 1996.

[9] Yeremei Aipin was not re-elected to the Peoples' Deputies of the Russian Parliament in the elections of May 1996, but remains the President of the Association of the Small Peoples of the North in Russia.

References

Current Digest of the Soviet Press. 1990. Native northern people eye their rights. 42(29):20-21.

Leninskaya Pravda. 1989. Programma assotsiatsii 'Spasenie Yugri' [Program of the association 'Salvation of Yugria']. 20 July.

The Lubicon Settlement Commission of Review: Final Report. Edmonton. 1993.

Sweetgrass. 1984. Allegations of genocide. (May/June):13.

Aipin, Yeremei. 1991. Not by oil alone; and USSR: Road of discord. *IWGIA Newsletter* 1(July/August): 36-39.

Balzer, Marjorie. 1983. Ethnicity without power: The Siberian Khanty in Soviet society. *Slavic Review* 42(4): 633-48.

Bond, Andrew R., Mark Bassin, Michael J. Bradshaw, George J., Leslie Dienes, Paul Goble, Gary Hausladen, Ronald D. Liebowitz, Philip R. Pryde, Lee Schwartz, Victor H. Winston. 1991. Panel on Siberia: Economic and territorial issues. *Soviet Geography* 32(6):363-432.

Bradshaw, Michael J. 1992. Siberia at a time of change: New vistas for western investment. *The Economist Intelligence Unit. Special Report.* London: Economist Intelligence Unit.

Campbell, Robert W. 1968. The economics of Soviet oil and gas. Baltimore, Maryland: The Johns Hopkins Press.

Chance, Norman A. and Elena N. Andreeva. 1995. Sustainability, equity and natural resource development in Northwest Siberia and Arctic Alaska. *Human Ecology* 23(2):217-240.

Churchill, Ward. 1988/1989. Last stand at Lubicon Lake: An assertion of indigenous sovereignty in North America. *IWGIA Document* 62:152-174.

Cloutier, Joe. 1988. Is history repeating itself at Lubicon Lake? *Canadian Journal of Native Education* 15(1):1-17.

Cornell, Stephen. 1988. *The return of the Native: American Indian political resurgence.* New York: Oxford University Press.

Dickason, Olive. 1992. *Canada's first nations: A history of founding peoples from earliest times.* Toronto: McClelland and Stewart.

Ebel, Robert E. 1994. *Energy choices in Russia.* Washington, D.C.: Center for Strategic and International Studies.

Elliot, Iain F. 1974. *The Soviet energy balance: Natural gas, other fossil fuels, and alternative power sources.* New York: Praeger Publishers.

Fleras, Augie and Jean Leonard Elliot. 1992. *The "nations within": Aboriginal-state relations in Canada, the United States, and New Zealand.* Toronto: Oxford University Press.

Forsyth, James. 1992. *A history of the peoples of Siberia: Russia's north Asian colony 1581-1990.* Cambridge: Cambridge University Press.

Goddard, John. 1991. *Last stand of the Lubicon Cree.* Vancouver, B.C.: Douglas and McIntyre.

Gorbachev, Mikhail S. 1995. Sibir-uskorennyi shag: Vystuplenie na soveshchanii partiino-khozyaistvennogo aktiva Tyumenskoi i Tomskoi oblastei [Siberia—intensified development: Speech at the meeting of party-economic leaders of Tyumen' and Tomsk regions]. Moscow: Izdatel'stvo politicheskoy literatury.

Gosudarstvennyi Komitet SSSR po Statistike. 1991. *Natsional'nyi sostav naseleniya SSSR po dannym vsesoyuznoi perepisi naseleniya 1989* [National Composition of the Population of the USSR according to the All-Union Population Census for 1989]. Moscow.

Gustafson, Thane. 1989. *Crisis amid plenty: The politics of Soviet energy under Brezhnev and Gorbachev*. Princeton: Princeton University Press.

IWGIA. 1990. *Indigenous Peoples of the Soviet North*. IWGIA Document No. 67 (July).

"L." (Anonymous). 1959. The new Soviet man. In *The Soviet crucible: Soviet government in theory and practice*. Samuel Hendel, ed. Princeton: D. Van Nostrand Company, Inc.

Laxer, James. 1983. *Oil and gas: Ottawa, the provinces and the petroleum industry*. Toronto: James Lorimer and Company.

Mihalisko, Kathleen. 1989. SOS for Native peoples of the Soviet North. *Report on the USSR*. (3 February): 3-6.

Nulty, Peter. 1992. The Black Gold Rush in Russia. *Fortune* (15 June): 126-130.

Osherenko, Gail. 1995. Indigenous political and property rights and economic/environmental reform in Northwest Siberia. *Post-Soviet Geography* 36(4): 225-237.

Petrunenko, Oksana. 1991. A Soviet reservation? Yes! *IWGIA Newsletter* 1(July /August): 40-41.

Pika, A. and B.B. Prokhorov. 1989. Soviet Union: The big problem of small ethnic groups. *IWGIA Newsletter* 57(May).

———. 1994. Neotraditsionalizm na Rossiyskom Severe i gosudarstvennaya regional'naya politika [Neotradionalism in the Russian North and state regional policy]. Moscow.

Postanovlenie: preziduma TsK profsoyuz rabochikh geologo-razvedochnykh rabot. 1963. *Protokol No. 25: Ob initsiative kollektivov burovyk brigad Tyumenskogo upravleniya t.t. Urusova S.N. i Tarasova, A. F.* [Decree: presidium of the CC trade union of geological workers. Protocol No 25: Concerning initiatives of their drilling brigades under the Tyumen' management comrades Urusova S.N. and Tarasova, A. F]. 19 March. Moscow.

Pratt, Larry and Ian Urquhart. 1994. *The last great forest: Japanese multinationals and Alberta's northern forests*. Edmonton: NeWest Press.

Prilozhenie No. 1 k postanovleniu TsK KPSS i Soveta Ministrov SSSR ot iulya 1956. *MEROPRIYATYA po dal'neishemy razvitiu khozyaistva i kul'tury v rayonakh Krainego Severa i otdalennykh mestnostyakh, priravnennykh k nim.* [Supplement No. 1 to decree of the CC of the CPSS and Union of Ministers of the USSR from July 1956. Measures on further development of the economy and culture in regions of the Far North and remote areas like them]. Fond 814, Opis' 1, Delo 4113, list 26-50.

Prokof'yeva, E.D., V.N. Chernetsov and N.F. Prytkova, 1956. The Khants and Mansi. In *The peoples of Siberia*, eds. M.G. Levin and L.P. Potapov. Chicago and London: The University of Chicago Press.

Riva, Joseph P. 1993. The petroleum resources of Russia and the Commonwealth of Independent States. In *The former Soviet Union in transition*. Richard F. Kaufman and John P. Hardt, eds. Armonk, N.Y.: M.E. Sharpe for the Joint Economic Committee Congress of the United States.

Schindler, Debra L. 1991. Theory, policy and the "Narody Severa." *Anthropological Quarterly* 64(2):68-80.

Slezkine, Yuri. 1992. From savages to citizens: The cultural revolution in the Soviet Far North, 1928-1938. *Slavic Review* 51(1): 52-76.

Smith, James, G. E. 1987. Canada: The Lubicon Lake Cree. *Cultural Survival Quarterly* 11(3): 61-62.

Tasch, Jeremy. 1996. Institutional morass and the challenge of development: Post-Soviet observations in fuel and environmental management. Paper presented at the 1996 Annual Meeting of the Association of American Geographers, Charlotte, North Carolina, 9-13 April.

Tollefson, Chris. 1996. Strategic lawsuits and environmental politics: Daishowa Inc. *v*. Friends of the Lubicon. *Journal of Canadian Studies* 31(1):119-132.

Yamal-Potomkam. 1991. Ustav: Assotsiatsii korennykh narodov Severa Yamalo-Nenetskii avtonomnogo okruga (respubliki) "Yamal-Potomkam!" [Charter: Association of indigenous northern peoples of the Yamal-Nenets Autonomous Okrug (republic) "Yamal for our progeny!"]. Salekhard. 31 January.

Zubek, Voytek. 1984. *Soviet industrial theory*. New York: Peter Lang.

4

Environmental Degradation and Indigenous Land Claims in Russia's North

Gail A. Fondahl

In the Russian North, several processes have constrained aboriginal peoples' abilities to continue to use and steward their traditional homelands. The original imposition of government by external (Tsarist and Soviet) institutions undermined local decision-making regarding resource management. Relocation policies during the Soviet period removed aboriginal persons from their traditional territories. Soviet officials also imposed new activities and new ways of pursuing old activities upon Native communities. All of these actions changed the ways indigenous people could exercise land tenure over their traditional territories. Competition from non-aboriginal uses (e.g., forestry, mining, hydrocarbon development, city construction) has also geographically circumscribed the spaces in which aboriginal peoples can pursue their chosen activities. As devastating as any of these processes to current and future prospects for aboriginal control over historic homelands is the environmental degradation that has accompanied the industrial and strategic development of the Russian North. Native peoples from the Kola Peninsula to Sakhalin Island have witnessed the desecration of the lands which sustain their cultural identity.

If all facets of aboriginal life have been influenced by environmental degradation, it is the traditional activities—reindeer herding, hunting, trapping, gathering and fishing—which have been most affected. It is also the land-extensive traditional activities which form the core of aboriginal identity. Thus, when environmental degradation affects the viability of these activities it attacks the very essence of aboriginal societies. This paper examines the nexus between traditional land-based activities, environmental degradation, and the current development of indigenous "land claims" in the

Russian North. After offering background on the general status of aboriginal peoples in the Russian North, it examines the importance of traditional activities—and the land that supports these—to the persistence of Native cultures. It then briefly cites evidence of environmental degradation's toll on traditional aboriginal activities in different areas of the Russian North. How these activities have become the focus for legislative reforms which address aboriginal rights is then examined. Challenges posed by ever more limited areas in which to practice traditional activities have in fact been a main stimulus of the legislation, which explicitly attempts to preserve a land base for the traditional activities. Yet a large gap remains between current legislation on aboriginal land rights and aboriginal groups' need for significant powers of self-determination over their homelands. In conclusion, I will raise the question of whether the focus on the environmental needs of traditional activities has circumscribed the terms of debate in which aboriginal peoples can engage regarding land claims at this point in time.

Indigenous Peoples of Northern Russia: Status at the End of the Millennium

Russia's Arctic and subarctic zones are home to well over two dozen distinct peoples (Map 4.1). These peoples represent a wide range of language families, from Saami (related to Finnish) to Eskimosy (related to the Alaskan Yup'ik) to Nanay (related to Manchurian). They also evince a varied set of adaptation to a zone which, stretching across 25 degrees of latitude and 160 degrees of longitude, offers a range of environments almost as varied as the peoples themselves. Individual groups and peoples have depended on reindeer herding, hunting and trapping of land and sea mammals, fishing, and trade. Most often, they employ a combination of these activities, seasonally shifting in emphasis, to dwell in the North.

Upon assuming power, the Soviet state identified the peoples of the North as exceedingly primitive, and in need of a special policy body to facilitate the transition to socialism (Sergeev 1995; Slezkine 1994). At the same time the Bolsheviks fingered the North as a storehouse of wealth to be exploited for the development of the new socialist state. In the first decade of Soviet power, planners deliberated on balancing aboriginal needs and state aspirations in debates regarding northern development policy, but by the mid-1930s, the latter took clear precedence over the former. When development concerns dictated, the state confiscated aboriginal lands[1] and relocated Natives.

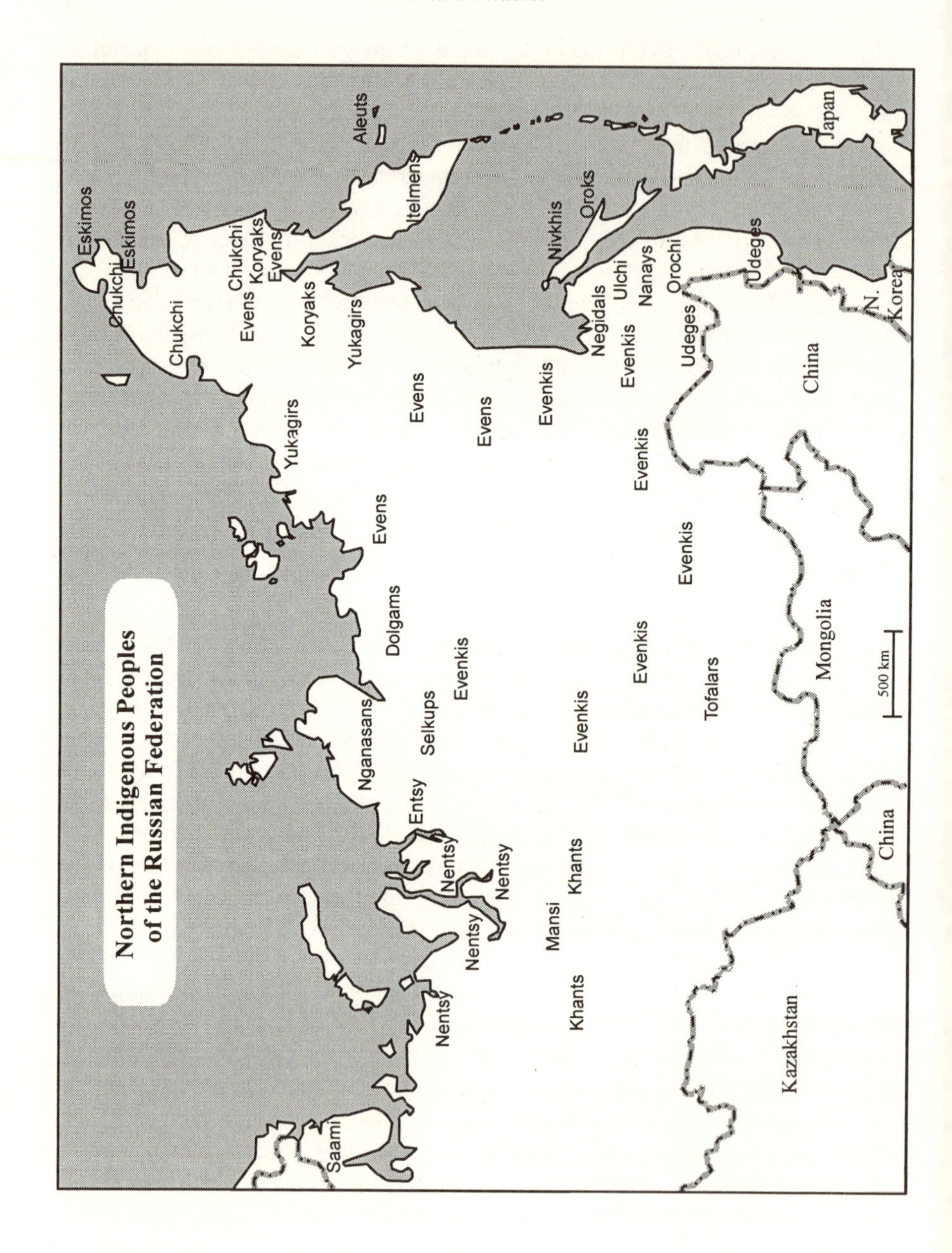
Northern Indigenous Peoples
of the Russian Federation
Eskimos
Eskimos
Chukchi
Chukchi
Chukchi
Koryaks
Evens
Evens
Koryaks
Aleuts
Itelmens
Yukagirs
Yukagirs
Evens
Evens
Evenkis
Nivkhis
Oroks
Negidals
Ulchi
Nanays
Orochi
Evenkis
Udeges
Udeges
Japan
N. Korea
China
Evenkis
Evens
Dolgans
Evenkis
Evenkis
Evenkis
Nganasans
Selkups
Evenkis
Tofalars
Mongolia
500 km
Entsy
Nentsy
Nentsy
Khants
Mansi
Nentsy
Nentsy
Khants
China
Kazakhstan
Saami

MAP 4.1

Given that the North was essentially closed to foreigners (and most Soviet citizens for that matter), outsiders knew little of the true situation which characterized aboriginal life in the Russian North. Frequent paeans to the Soviet system, often penned by members of these peoples, obscured their oppression (e.g., Volkova 1982). Indeed, a favorite genre of the Soviet press involved comparing the happy fate of Russia's indigenous northerners with the sorry fate of those in the capitalist states. Of the few Westerners who had a chance to visit Native villages, the most outspoken accepted this propagandistic view, naively or complicitly (e.g., Mowat 1970).

Following decades of such disinformation, *glasnost* revealed a situation all too familiar to the West. Like their counterparts in other circumpolar states, the indigenous peoples of Russia's northern areas have elevated rates of infant mortality, greater incidence of tuberculosis and a number of other diseases. They suffer from substantially higher suicide and homicide rates (Pika 1993; Zdorov'e 1996). Life expectancy runs a generation less than the perturbingly low Russian average (Pika and Prokhorov 1994). Living conditions in most Native villages remain squalid, with insufficient and thus overcrowded housing, lack of running water and indoor plumbing, etc. In general the northern Native peoples endure a lower level of well-being than that of the majority population. Unlike birth rates of the aboriginal populations of the North American North, which have contributed to high rates of growth overall (Knapp 1992), those among Russia's northern aboriginal peoples do not make up for the higher mortality rate, and growth of populations has been slow. In fact, six of the officially recognized northern peoples suffered absolute declines in numbers during one or both of the last two censuses [1970-79, 1979-89] (Osnovnie 1990).

Key to this social and physical malaise has been the destruction of cultural identity—a place-based identity tied to the land. Fifty years of Soviet nationality policy sought to sever those ties through assimilation of aboriginal northerners into mainstream Soviet society. Nomads were removed from their traditional territories and forcibly re-settled, children were reared in Russian boarding schools away from their parents and ancestral homelands. Individuals were encouraged to become modern industrial laborers, mobile and free from the constraints of place-basedness. Those remaining in traditional activities were to "rationalize" these fields in order to increase their productivity according to measurements stipulated by the state. "Rationalization" involved the incorporation of management strategies and technologies often developed in distant urban centers, and for other environments. Elders' land-based knowledge was devalued in the face of "scientific" research which indicated how to best modernize such traditional activities.

Assimilation has failed, both due to Russians' frequent unwillingness to accept aboriginal persons as equals, and to aboriginal persons' inability or unwillingness to complete the process of renouncing what can be characterized as a place-based identity. At the same time, the process has succeeded in undermining many aboriginal children's competency in their own cultures, in part by separating educational processes from place. Poorly trained for Russian-type industrial employment, which, in any event is often unavailable in their home villages, the children are no better prepared for taking up traditional activities. The latter require long apprenticeships in the field, alongside "professionals" experienced reindeer herders, hunters or fishers, in order to come to "know" the land (Anderson 1995).

Rationalization of traditional activities has also largely failed. Reindeer herding and hunting have suffered rather than profited from most attempts to "scientifically" alter management systems that have been carefully tailored to local environments over several centuries. Though statisticians eliminated most nomadism by creative counting (e.g., those who had houses were enumerated as "settled" even if they did not live in those houses), the herders and hunters still must remain on the land, at a distance from the villages. Snowmobiles and mobile housing units, for example, have offered a minority of indigenous peoples easier living and working conditions. Unfortunately, more capital has gone into poorly thought-out rationalization initiatives such as inappropriate pasture fencing programs, rather than useful technology.

The lack of Soviet authorities' respect for the traditional land-based knowledge which guided the development of herding, hunting and fishing over the past centuries has, *inter alia*, translated into a declining prestige of traditional practices with Native communities themselves. Yet aboriginal value systems, where they maintain even a compromised vitality, can confound attempts of Native persons to adapt to Russian industrial society. This dilemma, much studied during the late Soviet period (Boiko 1986, Boiko et al. 1987, Boiko and Popkov 1987, Boiko 1988), has greatly disturbed older generations of aboriginal societies. Parents see their children caught between two cultures, destined to live at the bottom rung of the Russian culture, in the least prestigious, unskilled physical labor jobs available in the village, or in poorly supported traditional activities. Many people, indigenous and others, consider the continuation of such traditional activities to be absolutely critical to the persistence of indigenous cultures, and advocate support for the development of these. They also note the nexus of these activities' continuation and environmental degradation.

Traditional Activities, Aboriginal Identities and Land

> *Destroy our reindeer breeding and our traditional lands and you destroy us, the Even, as a people. (V.A. Robbek)*[2]

The traditional activities of northern aboriginal peoples have indeed long served as the distinguishing emblems of these peoples within a Soviet context of ethnic homogenization. Northern peoples differed from other citizens of the Russian Federation due to their involvement in activities that required an intimate connection with, and an extensive use of, expansive homelands. If symbolic of primitivism in the eyes of many Soviet citizens, the traditional activities also symbolized a special, harmonic and intense interaction with the natural environment. "Children of nature," an epithet frequently used by the non-indigenous population, is also employed by indigenous persons themselves to underscore this imagined special relation.[3] Interactions with their environment, via the pursuit of "traditional" activities, was fundamental to their physical, mental, psychological, spiritual—and thus cultural—well-being.

Viewed perfunctorily as economic pursuits by Soviet officials, traditional land-based activities incorporated social practices, supported language preservation, and sustained elements of spiritual beliefs and value systems through five decades of Soviet attempts at assimilation. For instance, reindeer husbandry provided milk, transport, skins and meat, all of which could be (and was) commercialized and co-opted into a larger state economic system. But among many of the reindeer-herding peoples, the deer also conveyed a family's protective spirits, provided not only physical but spiritual nourishment at life-event celebrations, and accompanied the owner on her or his voyage from this world to the next. A rich indigenous vocabulary surrounding reindeer husbandry had no corollary terms in Russian; where non-Native workers entered this field, they had to learn the Native language if they hoped to understand and express finer nuances of "the trade." In fact, Soviet planners charged with "rationalizing" traditional activities discovered to their chagrin that they could not disentangle the social and spiritual from the economic in such activities, a factor which hindered their "modernization."

The indigenous northerners' traditional activities require areally extensive territories—and an areally extensive way of life, characterized by nomadism. In doing so, these activities play a critical role in perpetuating aboriginal ties to land. "Home" for nomadic peoples often constituted a territory rather than a dwelling. As noted for the aboriginal population of the Katanga Region (Irkutsk Province), the Evenki:

> Evenkis psychologically identify themselves with a home, conceiving of it not in the form of an abstract tipi in an abstract place, but in the form of that tayga territory where their hunting, pasturing, and fishing grounds, with the material objects of subsistence, are located, calling this territory "my land," "home." (Sirina 1992:80)

Such territories, the parameters of which were determined in part by the ecological requirements of the traditional activities, in turn spatially define the identity of groups comprising the different northern indigenous nations. If the natural environment of one's territory provides instrumental resources which support one's physical needs, "known" territory itself serves as well as a symbolic resource which mediates the definition of Self in relation to Other (Butz 1996; Anderson 1995). In a post-Soviet environment allowing the renegotiation of aboriginal-state relations, indigenous northerners have sought to reassert their distinct identities, including their territorial component. They have done so through demands for legal rights over homelands and legal protection from the further environmental degradation of these territories. Many factors challenge this reassertion, not the least of which is the scale of environmental degradation to date.

Traditional Activities and Environmental Degradation: a Brief Geographical Survey

> *Our Native lands are being annexed and barbarically destroyed by rapacious petroleum and natural gas, coal, gold and non-ferrous mining interests without any form of just compensation. . . . and this phenomena [sic] is depriving us of our lands and rights to part of the resource wealth, [and] deprives us of our basic right—a right to life. (Social . . . 1996)*

Since the 1930s, the Russian North has witnessed a rapid growth in industrial development, with few precautions taken to protect the subarctic and Arctic environment. Circumpolar environments are extremely vulnerable to disruption, and, once disturbed, are slow to recover. Industrial activities have removed millions of hectares from use by indigenous people, and compromised the productivity of yet millions more. Activities directly linked with such industrial activities, such as the recreational hunting (legal and illegal) by laborers who came north to "build socialism," have further threatened the viability of traditional activities and the land base on which they depend.

Environmental degradation in the North may be broken into several categories: chemical pollution of air, land and water; mechanical disturbance of flora, land and water; thermal pollution of these media, and biodepletion of fauna and flora through overharvesting are the main villains. Air pollution, especially associated with large-scale mineral processing plants, such as those of the Taymyr and Kola Peninsulas, has, through acid precipitation, brought about the decimation of thousands of square kilometers of tundra and taiga (Vilchek et al. 1996). Heightened concentrations of sulfur, copper, nickel and cobalt in soils, water and lichens also stem from these sources. Chemical pollution of water and land is especially notable in the oil and gas bearing regions of northwestern Siberia, where poor drilling practices and breaks in the pipeline transport systems contribute millions of tonnes of hydrocarbons to the environment every year.

Mechanical disturbances include the indiscriminate use of off-road vehicular traffic over fragile tundra and taiga ecosystems, and the thermokarst processes which result from agitating the active layer which insulates underlying permafrost; the silting of rivers during construction, transport, timber operations, and other activities; and the clear-cutting of forests. Flooding vast areas of taiga and tundra in the construction of hydroprojects also may be listed under this category. Such activities, in addition to being pollution-rich in particulate matter which settles out, cause changes in surface albedo. Such changes can, through feedback mechanisms, be amplified and initiate further changes.

Thermal pollution of land and rivers, through industrial effluent, the creation of artificial bodies of water, and the disturbance of permafrost regimes, also compromises northern environments. All the processes mentioned above remove from use, or jeopardize the productivity of, fishing habitat, hunting grounds, marine-mammal harvesting areas and reindeer pasture. Adding to these challenges to traditional activities, migrant laborers have contributed to the biodepletion of many wild game and fur species, through indiscriminate hunting, as well as to the decimation of some domesticated reindeer stocks, and the overharvesting of some plant species (Syroechkovskiy 1989; Vilchek et al. 1996).

Estimates of damage to northern ecosystems vary widely, but suggestive are the following statistics: damage to Arctic vegetation by pollution has affected 50,000 km^2; mechanical disturbances have altered 31,000 km^2; deforestation has occurred over some 150,000 km^2, and fires have decimated 180,000 km^2 of Arctic and northern ecosystems. Overgrazing of reindeer pastures affects some 500,000 km^2 (Vilchek et al. 1996:33-37). Spatially uneven, the environmental degradation has devastated some areas of the Russian North, damaged others, and been relatively absent from yet others. Indigenous peoples in the most affected areas of the Russian North face very real threats to their abilities to pursue land-based activities. We must contextualize environmental degradation as but one of the many challenges facing these peoples, and thus realize that even modest levels of degradation, in concert with other social, economic and cultural depredations suffered by indigenous groups, can prove catastrophic to the continuation of traditional land-based activities.

Traditional Activities as the Focus for Legislative Reform

> *In the majority of regions inhabited by [the numerically Small Peoples of the North] the ecological situation has sharply intensified, the systematic destruction of established norms and rules of natural resource use has been allowed.* (O dopolnitel'nykh 1991)

During the Soviet period, environmental regulations did little to protect traditional aboriginal activities, and the northern environments on which they depend, from the onslaught of environmental threats briefly catalogued above. As soon as Gorbachev's glasnost policies allowed criticism of the system, indigenous peoples underscored the danger environmental degradation posed to the continuance of traditional activities and called for the protection of the remaining intact lands, compensation for those already ruined, and for legislation that would return control over the way northern lands were used to the indigenous population (*Materialy* 1990).

Russia's lawmakers have responded to a persistent indigenous lobby with a host of legislation. Some of the federal decrees [*postanovleniya*], presidential edicts [*ukazy*] and federal laws [*zakony*] specifically address aboriginal peoples' rights to a land base. Other, more general legislation, especially that on resources, incorporates articles stipulating such protection. Perhaps most notably, the new (1993) Russian Federation Constitution requires the "defense of the age-old living spaces and traditional way of life of numerically small ethnic communities (Konstitutsiya 1993:§72).

Federal legislation requires the maintenance of the environment in areas where numerically Small Peoples live (O svobodnom 1990, §5), the establishment of protected territories in these areas (Zemelnii 1991, §15; O dopolnitel'nykh 1991, §7; O sotsial'no-ekonomicheskom 1992, §3; O neotlozhnykh 1992, §1), and the allocation of land grants to indigenous persons wishing to pursue traditional activities (Ob uporyadochenii 1992). Legislative initiatives also give indigenous peoples priority rights to resource usage connected with traditional activities (Zemelnii 1991, §28; O dopolnitel'nykh 1991, §§1,7) and eases the pursuance of these activities through tax breaks and favorable land rents and use of otherwise restricted lands for subsistence use (Zemel'nii 1991, §§51,90; O neotlozhnykh 1992, §1). Environmental rehabilitation is called for (Gosudarstvennaya 1991, §29), as is minimizing future environmental disturbance of lands used by indigenous peoples (O sotsial'no-ekonomicheskom 1992, §§6,12; Osnovy 1993, §51).

The emphasis of the recent legislation is on the establishment and protection of a land base for traditional aboriginal activities. Two approaches that could contribute to protecting against environmental degradation can be identified in the legislative acts addressing indigenous rights to pursue traditional activities; those summoning the allocation of land allotments to indigenous *obshchinas*[4] which are primarily engaged in traditional activities; and those requiring the demarcation of protected territories in which activities other than traditional ones are severely restricted.

These two approaches in fact serve as the backbone of aboriginal "land claims" in the Russian North (Fondahl 1996). Each approach has inherent limitations, as discussed below. The greatest shortcoming at present stems from the weakness of implementation mechanisms: provincial and republican governments may choose to activate or block such reforms, given the absence of federal laws (both approaches are outlined in federal decrees and edicts). Where republics and provinces have begun to implement such reforms, they have done so selectively. In the following sections, the two approaches are described, and their ability to address environmental protection is assessed.

Land Allocations to Indigenous Obshchinas

In accordance with several federal decrees and edicts, an indigenous collective of individuals or families—an obshchina—may petition the *raion* (county) in which it resides for an allotment of land on which to pursue traditional activities (Gosudarstvennaya 1991, O neotlozhnykh 1992). It does not receive title to the land. Rather, it receives rights to use the renewable resources related to traditional activities: the reindeer pasture, game animals, fur animals, fish, edible and medicinal plants. Timber cutting is limited to subsistence use, whereas the other resources noted above may be pursued commercially as well as for domestic usage. A presidential edict called for land to be allotted in perpetuity, with rights of inheritance by obshchina members; in practice such allotments have been made for finite periods, ranging from one to twenty-five years (O neotlozhnykh 1992; Fondahl 1995).

For many of the indigenous northerners the obshchina served as the basic indigenous territorial unit during the pre-Soviet period (Vakhtin 1993,

Pika and Prokhorov 1994). By identifying the obshchina as the rightful unit of property management the new Russian Federation legislation acknowledges the environmental and cultural sustainability of the traditional land-tenure systems of indigenous peoples. Land reform in general has been based on a belief that investing responsibility for property in individuals would improve both the productivity and sustainability of activities supported by that property. In this particular case, a strong indigenous lobby convinced that government that "privatization" of lands was best replaced by communal indigenous management. The issue of federal vs. indigenous communal ownership, a thorny one, is beyond the scope of, and not critical to, this paper.

Between April 1992, when the presidential edict was declared, and February 1997, some 2,300 obshchinas registered with local officials. However, a large part of these have since dissolved. Many only received a land allotment for a season or year, contrary to the edict's call for land grants "in perpetuity, with rights to inheritance." (O. Murashko, pers. comm., 23 February 1997). Without land, such obshchinas could not persist. For those who have received land, allotments range in size from thousands to hundreds of thousands of hectares, some offering credible basis for survival, others not.

Obshchina territories do not provide a comprehensive solution to either indigenous land rights or environmental protection of Native homelands. Based as they are on economic activities, they fail to offer an assured means to exert control over important spiritual and other sites. Neither do the obshchinas enjoy full exclusionary rights over the land allotted to them. County administrations in which they are located may grant licenses for subsurface prospecting. Forming an archipelago of land allotments across the Russian North, the islands of "traditional" land use can be easily contaminated by activities outside their boundaries. Upriver mining, nearby forestry operations, and a host of other activities may undermine the ability of an obshchina to viably develop hunting or reindeer herding on the land allotted to it for this purpose. To illustrate such challenges, in one particularly appalling case, an obshchina in the Chita Province received a territory within a military test range, an area largely fouled and burnt from forty years of weapons testing (Author's field notes, 1994; Aruneev 1992).

Moreover, the very spatial structure of late twentieth century obshchinas contravenes characteristics of their antecedents which encouraged sustainable resource use and limited overgrazing or overharvesting. Boundaries of traditional aboriginal obshchinas appear to have been flexible and permeable, recognized but manipulated as required by game numbers and distributions and human needs. When environmental conditions required, members of one obshchina might approach a neighboring group to ask permission to use this group's territory until recovered. An indigenous former herder- and hunter-turned-political-leader noted concerns about the spatial constraints today's obshchina territories seem to imply: "Naturally, to put reindeer with their master in such a cage will give nothing but the death of the deer" (N.A., Interview, August 1995). Obshchina boundaries, as currently drawn by local Committees for Land Reform, appear more rigidly construed, in keeping with state delineation of all kinds. Will indigenous groups will be able to work to overcome these imaginary lines on the landscape to ensure sustainable stewardship of resources across territories? To the extent that natural resource use on obshchinas is regulated by legislation (federal, republican and provincial), indigenous peoples are constrained in (re)developing their own mechanisms for avoiding internal pressures of environmental degradation.

Protected Territories

Indigenous peoples have stressed the need for protection of larger territories than those afforded by the obshchina allotments. Early in the period of glasnost, one Khanty summarized the views of many, calling for "zones of life" (Aipin 1989). A number of legislative decrees and a presidential edict have called for the creation of a variety of protected territories along these lines: "zones of priority nature use" [*zoni prioritetnogo prirodopol'zovanie*], "reserve zones" [*zapovednye zoni*], ethnocultural parks [*ethnokul'turnye parki*] and "territories of traditional nature use" [*territorii traditsionnogo prirodopol'zovaniya*, henceforth TTPs] (O dopolnitelnykh 1991, O neotlozhnykh 1992). Most developed in both concept and legal basis is the TTP. As of February 1997, a draft law on such formations still awaited passage (O. Murashko, pers. comm., 23 February 1997).

TTPs are areas of substantial size from which industrial activities are excluded, but traditional activities allowed. Alienation of lands within the TTPs is supposed to be permitted only by referendum of the local indigenous population. Thus, the TTPs allegedly provide a relatively secure land base on which to continue pursuit of traditional activities, while allowing indigenous peoples who wish to develop industrial sources of income to do so. This ability to remove land for industrial development, it is worth noting, conforms with international conventions on indigenous self-determination.

Throughout much of Siberia, provincial leaders have ignored the legislation that stipulates the establishment of TTPs. They cite the absence of a law requiring such TTPs, and hold that they will await action until such a law appears. In a few areas of the Russian North such TTPs have been created, with interim management measures developed to serve until a law is passed (Mikheev 1995).

TTPs provide a more comprehensive protection for traditional activities, and specifically the land base which they require, than do obshchina territories, mainly because of their potential size. Obshchina territories in fact can be allocated lands within TTPs, providing additional protection for the traditional activities developed by the obshchina.

The TTP have been heralded as a bargaining chip in state-aboriginal negotiations that pit the need for industrial development in the North against the need to protect a significant land base for the perpetuation of the culturally important traditional activities for the aboriginal population. Under a scenario based on land claims settlements in other parts of the Circumpolar North, indigenous peoples would extinguish their "nominal and very undefined rights to 'historical territories'" in return for "real rights, legislatively confirmed and guaranteed by the government" over the territories of traditional nature use" (Pika and Prokhorov 1994:91). Such a strategy mirrors that used in the Alaska Native Claims Settlement Act (1971), the James Bay Northern Quebec Act (1975) and the Inuvialuit Final Agreement (1984). Governments of northern administrative-territorial units (e.g., *okrugs, oblast's*) might even be willing to help finance the formation of such territories within their borders in exchange for indigenous peoples surrendering their claims to natural resources outside of these borders (Pika and Prokhorov 1994:91,92).

A case study of an integrated system of three such adjoining TTPs, created in the three northern counties of the Chita Oblast', suggests several limitations (Fondahl forthcoming; see Mikheev 1995). In terms of alienation of lands for industrial development, it appears that it will not necessarily be the indigenous inhabitants who decide via a referendum: in this case, all persons living within the county in which a TTP has been created may vote in such a referendum. Within Northern Chita, as within much of the north, the bulk of the population is non-Native, of relatively recent origin, and can attribute its arrival to opportunities in the non-renewable resource extraction industries. To what extent will individuals in this cohort vote for preservation over opening up new resources to exploitation, when that may mean job opportunities?

Other issues plague the nature of Northern Chita's TTPs as ecological reservoirs for the protection of traditional activities. Limited to mainly alpine tundra, the TTPs protect the upper reaches of many watersheds, and important reindeer-herding habitat, but very little land valuable for hunting, gathering of many important forest plants, or fishing. Federal legislation underscores the importance of providing support to the *integrated complex* of traditional activities, recognizing that few indigenous peoples depended on a single activity alone for their livelihoods (O neotlozhnykh 1992). The TTPs as currently configured fail to offer such support. They also fail to protect key spiritual sites within the area, a limitation that a number of local indigenous persons consider objectionable (Author's field notes, 1994).

Aboriginal Land Claims: Constraints and Inadequacies

Legislation which addresses the persistence of aboriginal traditional activities and the devastation of the environment has begun to be adopted. Yet several inadequacies plague the current legislative, including deficiencies in the status of legislative initiatives, shortcomings in the concepts of protected territories, and a myopia regarding the focus on traditional activities.

Many of the provisions regarding establishment of obshchina allotments, TTPs, and other territories designating for protection for traditional activities, called for in decrees and edicts have not yet been incorporated into laws. While the same decrees and edicts summoned the quick passage of laws, such laws still await adoption. Decrees and edicts do not carry the weight of laws. The approach of some local governments has been to ignore decrees and edicts; they declare that they will await action on establishing such "Native territories" until the passage of laws requires such. Others have moved on the decrees and edicts by creating their own temporary provisions, until federal laws are adopted. Such a tactic can be (and has been) used to subvert, rather than uphold, the intent of federal decrees and edicts. Where contradictions require resolution, at present their settlement depends too often on the persuasion of local administrators.

The single most important law regarding traditional indigenous activities and their protection, the comprehensive Law on the Legal Status of Numerically Small Peoples of the North, has languished in Russian parliamentary chambers for several years (Murashko 1996). It has suffered sequential revisions, each diluting its powers. In 1995, after the law twice received the federal parliament's approval, President Yeltsin refused to sign it, sending it back for further revision (dilution). Other laws critical to the establishment of a recognized land base for traditional activities have not fared better.

Moreover, legislation on indigenous rights to the land and resources necessary to pursue traditional activities contradicts other laws and decrees passed by the federal government. On occasion it also comes into conflict with laws and decrees adopted by the republics of the Russian Federation (e.g., Sakha, Komi, Buryatia) regarding indigenous rights. For instance, different decrees on indigenous land allotments stipulate different terms for such allocations, ranging from a few years to lifelong rights with the legal ability to pass such land allotments on to one's descendants. Likewise, while some decrees call for such allotments to be made to indigenous peoples without charge, other laws require rents to be paid to forest district offices Such contradictions require resolution; at present their settlement depends on the convictions of local administrators.

At another level, the concept of protected territories for aboriginal activities itself poses threats to aboriginal land rights, if offering promise for environmental protection of land critical to the survival of traditional activities. To date, the establishment of most protected territories has signified a juridical affirmation of state ownership of such lands—and circumscribed the activities which the local population, including indigenous persons, can undertake on such lands. The TTPs could offer some room for negotiation, given the stipulation allowing for alienation of lands from these territories once they have been established. As intimated above, such stipulations only enhance aboriginal rights if the indigenous population itself determines whether such alienation occur. Indigenous peoples in the mid-1990s have been involved to a greater decree in the delineation of protected territories (K.Klokov, pers. comm., 27 May 1996; P. Prokosh, pers. comm., 12 July 1996). However, their role in the continued management of resource use, traditional and otherwise, within protected areas, remains largely ill-defined. Co-management regimes of the kind increasingly employed in the North American North (Notzke 1995) have not developed in the Russian North. Protected territories too often offer more restriction than protection for indigenous peoples who increasingly rely on livelihoods based on a mix of traditional and non-traditional activities.[5]

Indeed, the focus on traditional activities itself may hinder the comprehensive development of aboriginal land rights. The criticality of these traditional activities for indigenous cultural survival cannot be overemphasized. Yet, by constricting the dimensions of land claims to lands for traditional territories, the state dispossesses aboriginal peoples of the ability to creatively combine these activities with non-traditional activities, and thus to develop economically viable *and* flexible strategies within their homelands. Aboriginal rights based on a set of traditional activities as defined by the state limits options. It may also hinder the potential for sustainably developing non-traditional activities (cf. Osherenko 1995). While there is no absolute assurance that indigenous peoples managing or having significant input into the management of non-traditional activities will encourage those activities to be practiced in a more sustainable manner, the persistence of a stewardship ethic founded on intimate ties to the land suggest the potential for doing so. More colorfully asserted:

> If I'm standing around someone else's office, I might think nothing of spitting on the floor. It's the cleaning woman's job to clean it, it's not my floor. But would I ever think of spitting on the floor of my own home? If a person controls the land, he will take care of it. (Author's fieldnotes, Chita Province, June 1994)

In this transitional economic and political period, pressures for industrial development in the North will increase, by both domestic and foreign developers. Vesting rights to substantial areas of the North in the aboriginal population will likely improve the potential for the development of both traditional activities, and the possibilities for more rationale development of non-traditional activities. Until legislation provides for such comprehensive land claims, the chances for both promoting cultural survival and reducing environmental threats remain imperiled.

---

[1] Land was officially nationalized in 1917. However, the *de facto* alienation of land from indigneous users took place gradually; at the end of the Soviet period, in some corners of the North control over land had only been partially ceded by indigenous peoples, while throughout much of the North, their authority was undermined to a much greater degree.

[2] Statement made by V.A. Robbek, Director, Institute of the Problems of Northern Minorities, Yakutsk, Sakha Republic (Yakutia), at the Second International Working Seminar on the Problems of Northern Peoples, 26 May 1996, Prince George, B.C.

[3] Krupnik (1993) offers an excellent argument challenging this simplistic view of "aboriginal peoples in tune with nature." Here, the critical point is that of *imagined* harmony as defining characteristic of the various Peoples of the North, by both the peoples themselves and other Russian citizens (cf. Anderson 1983).

[4] *Obshchina* in the ethnographic literature refers to a small, kin-based indigenous community at a sub-clan level. Ethnographers have identified the obshchina as the basic social-territorial unit of organization for most of the peoples of the Russian North (Pika and Prokhorov 1994). Recent draft legislation has offered a modified definition: the obshchina is "a voluntary unification of families and properties from among [indigenous] rural inhabitants for collaborative development in the management of deer pastures and hunting and fishing grounds" (Draft Law n.d., §6).

[5] See Hugh Beach's article in this volume for further discussion of this general theme.

References

Anderson, B. 1983. *Imagined communities: Reflections on the origin and spread of nationalism.* Revised Edition: London: Verso.

Anderson, D.G. 1995. National identity and belonging in Arctic Siberia: An ethnography of Evenkis and Dolgans at Khantaiskoe Ozero in the

Taymyr Autonomous District. Ph.D. Dissertation, University of Cambridge.

Aruneev, G. 1992. [No title recorded.] *Sovetskiy Sever* (Newspaper of Tungiro-Olëkma County) 27 February 1992.

Boiko, V.I. 1988. *Sotsial'no-ekonomicheskoe razvitie narodnostey Severa. Programma koordinatsii issledovaniy* [Socio-economic development of the Peoples of the North. A program for the coordination of research]. Novosibirsk: Nauka.

———. 1986. *Sotsial'nye problemi truda u narodnostey Severa* [Social problems of labor among the Peoples of the North]. Novosibirsk: Nauka.

Boiko, V.I., Yu. P. Nikitin, A.I. Solomakha. 1987. *Problemy sovremennogo sotsial'nogo razvitiya narodnostey Severa* [Problems of contemporary social development of the People's of the North]. Novosibirsk: Nauka.

Boiko, V.I. and Yu. V. Popkov. 1987. *Razvitie otnosheniya k trudu u narodnostey Severa pri sotsializme* [Development of relations toward labor among the Peoples of the North under socialism]. Novosibirsk: Nauka

Draft Law. n.d. Draft law of the Russian Federation. Foundations of the legal status of indigenous peoples of the Russian North. Typescript.

Fondahl, G. forthcoming. Freezing the frontier? Territories of traditional nature use in the Russian North, *Progress in Planning.*

———. 1995. Legacies of territorial reorganization for indigenous land claims in northern Russia, *Polar Geography and Geology* 19(1):1-21.

Gosudarstvennaya. 1991. Gosudarstvennaya programma razvitiya ekonomiki i kul'tury malochislennykh narodov Severa v 1991-1995 godakh. Sovershenstvovaniye natsional'no-territorial'nykh obrazovaniy i ikh yuridicheskogo statusa [State program for the development of the economy and culture of the numerically Small Peoples of the North in 1991-1995. Improvement of national-territorial formations and their juridical status]. Postanovelnie Kabineta Ministrov SSSR i Soveta Ministrov RSFSR N°84, 11 March 1991.

Knapp, G. 1992. *The population of the Circumpolar North.* Anchorage: Institute of Social and Economic Research, University of Alaska.

Konstitutsiya. 1993. *Konstitutsiya Rossiyskoy Federatsii, prinyata vsenarodnym golosovaniem 12 dekabrya 1993 g.* [Constitution of the Russian Federation, adopted by a nationwide plebiscite, 12 December 1993.] Moscow: Yuridicheskaya Literatura.

Krupnik, I. 1993. *Arctic adaptations: Native whalers and reindeer herders of northern Eurasia.* Marcia Levenson, trans. and ed. Hanover, N.H.: University Press of New England.

Materialy. 1990. *Materialy s"ezda malochislennykh narodov Severa* [Materials of the Congress of Numerically Small Peoples of the North]. Moscow.

Mikheev, V.S., ed. 1995. *Traditsionnoe prirodopol'zovanie Evenkov. Obosnovanie territoriy v Chitinskoy Oblasti* [Traditional nature use of the Evenkis. Bases for territory in the Chita Province]. Novosibirsk: Nauka.

Mowat, F. 1970. *The Siberians.* New York: Penguin Books Inc.

Murashko, O. 1996. Legal status of the indigenous numerically Small Peoples of Russia, *Zhivaya Arktika* 3:11-12.

Notzke, C. 1995. A new perspective in aboriginal natural resource management: Co-management, *Geoforum* 26(2):187-209.

O dopolnitel'nykh. 1991. O dopolnitelnykh merakh po ulushcheniyu sotsialno-ekonomicheskikh usloviyakh zhizni malo-chislennykh narodov Severa na 1991-1995 gody [On additional measures to improve the socio-economic conditions of life of the numerically Small Peoples of the North in 1991-1995]. *Postanovlenie Kabinetov Ministrov SSSR i Sovet RSFSR N°84*, 11 March 1991.

O nedrakh. 1992. O nedrakh [On sub-surface resources]. Zakon RSFSR ot 21 fevralya 1992.

O neotlozhnykh. 1992. O neotlozhnykh merakh po zaschte mest prozhivaniya i khozyaystvennoy deyatel'nosti malo-cheslennykh narodov Severa [On urgent measures for the protection of the places of residence and economic activity of the numerically Small Peoples of the North]. *Ukaz Prezidenta Rossiyskoy Federatsii N°397*, April 22, 1992.

O sokhranenii. 1993. O sokhranenii prirodnogo kompleksa sredy prozhivaniya Udegeytsev, Nanaytsev i Orochey v Pozharskom Rayone Primorskogo Kraya [On maintaining the natural complex of the environment inhabited by the Udege, Nany and Oroche in the Posharskom *Raion* of the Primorskiy Krai]. Postanovlenie Soveta Natsional'nostey Verkhovnogo Soveta Rossiyskoy Federatsii ot 24 fevralya 1993.

O sotsial'no-ekonomicheskom polozhenii. 1992. O sotsial'no-ekonomicheskom polozhenii rayonov Severa i priravnykh k nim mestnostey [On the socio-economic situation of the raions of the North and places equated with them]. Postanovlenie S"ezda narodnykh deputatov Rossiyskoy Federatsii ot 21 aprelya 1992.

O sotsial'no-ekonomicheskom razvitii. 1992. O sostial'no-ekonomicheskom razvitii rayonov Severa i priravnennykh k nim mestnostey. [On the socio-economic development of the raions of the North and places equated with them]. Postanovlenie Pravitel'stvo Rossiyskoy Federatsii ot 12 maya 1992 g. No. 308.

O svobodnom. 1990. O svobodnom natsional'nom razvitii grazhdan SSSR, prozhivayushchikh za predelami svoikh natsioal'nykh gosudarstvennykh obrazovaniy ili ne imeyushchikh na territorii SSSR [On the free national development of citizens of the USSR, living beyond the boundaries of their national state formations or not possessing such in the territory in the USSR]. *Zakon SSSR*, April 26, 1990.

O vypolnenii. 1993. O vypolnenii Postanovleniya Verkhovnogo Soveta Rossiyskoy Federatsii ot 15 yulya 1992 goda "O neotlozhnykh merakh po stabilizatsii sotsial'no-ekonomicheskogo polozheniya v Chukotskom avtonomnom okruge i v Magadanskoy oblasti" [On fulfilling the decree of the Supreme Soviet of the Russian Federation of 15 July 1992 "On urgent measures to stabilize the socio-economic situation in the Chukotka Autonomous Okrug and in the Magadan Oblast']. Postanovlenie Soveta Natsional'nostey Verkhovnogo Soveta Rossiyskoy Federatsii ot 29 yunya 1993.

Ob uporyadochnii. 1992. Ob uporyadochenii polzovaniya zemelnykh uchastkami, zanyatami pod rodovye, obshchinnye i semeynye ugodya malochislennykh narodov Severa [On regulations for the use of land plots occupied by clan, obshchina and family lands of the Numerically Small Peoples]. Decree of the Prezidium of the Supreme Soviet of the Russian Federation N°2612-1, 30 March 1992.

Osherenko, G. 1995. Property rights and transformation in Russia: Institutional changes in the Far North, *Europe-Asia Studies* 47(7):

Osnovnye. 1990. *Osnovnye pokazateli razvitiya ekonomiki i kul'tury malochislennykh narodov Severa (1980-1989 gody)* [Basic indices of development of economy and culture of the Numerically Small Peoples of the North]. Moscow.

Osnovy Lesnogo. 1993. Osnovy Lesnogo zakonodatel'stva Rossiyskoy Federatsii [Foundations of the forest legislation of the Russian Federation]. Prinyati Verkovnym Sovetom Rossisyskoy Federatsii 6 marta 1993 g.

Pika, A. 1993. The spatial-temporal dynamic of violent death among the Native peoples of Northern Russia, *Arctic Anthropology* 30(3):61-76.

Pika, A.I., and B.B. Prokhorov. 1988. The big problems of the Small Peoples. *Kommunist* 16:76-83. (Translated into English: Soviet Union: The big problems of small ethnic groups, *IWGIA Newsletter* 57:123-135.)

Pika, A.I., and B.B. Prokhorov, eds. 1994. *Neotraditsionalizm na Rossiyskom Severe* [Neotraditionalisn in the Russian North]. Moscow.

Sergeev, M.A. 1955. *Nekapitalisitcheskiy put' razvitiya malykh narodov Severa* [The non-capitalist path of development of the Small Peoples of the North]. Trudy AN Instituta Etnografii, NS, Volume 27.

Slezkine, Yu. 1994. *Arctic mirrors: Russia and the Small Peoples of the North.* Ithaca: Cornell University Press.

Social Organization and Movements of Indigenous Peoples of the North. 1996. Discrimination against indigneous people of the North in the Russian Federation. A statement. 4 March 1996. (Circulated on Internet at request of V.B. Shustov, General Secretary, Association of Indigenous Peoples of the North, Siberia and Far East.)

Syroechkovskiy, E.E. 1989. *Kraynii Sever: Problemy okhrany prirody* [The extreme North: Problems of nature protection]. Nauki o Zemli 8, Moscow: Izdatel'stvo "Znanie"

Vakhtin, N.B. 1994. Native Peoples of the Russian Far North. In: *Polar peoples: Self-determination and development.* Minority Rights Group (eds.), 29-80. London: Minority Rights Group.

———. 1993. Indigenous people of the Russian Far North: Land rights and the environment. Paper to be discussed at the United Nations Consultations on Indigenous Peoples. July 1993 (Typescript).

Vilchek, G.E., T.M. Krasovskaya, A.V. Tsyban and V.V. Chelyukanov. 1996. The environment in the Russian Arctic: A status report, *Polar Geography* 20(1):20-43.

Volkova, K.V., comp. 1982. *Bratstvo.* Novosibirsk: Zapadno-Sibirskoe knizhnoe izdatel'stvo.

Zdorov'e. 1996. Zdorov'e khozyaev Severa [Health of the owners of the North]. *Informatsionnii Byulleten' N°1*, Informatsionnii Tsentr Korennykh Narodov Rossii "L'auravetl'an," 3-4. (Excerpted from *Vestnik "Meditsina dlya Vas* 14(21), July 1996.)

Zemel'nii. 1991. Zemel'nii Kodeks RSFSR [Land codex of the RSFSR]. Prinyat 25 aprelya 1991.

5

A Survey of Pollution Problems in the Soviet and Post-Soviet Russian North

Craig ZumBrunnen

This chapter takes a broad perspective on environmental pollution problems in the Russian North, with the goal of supplementing the analyses of global problems of the Arctic Sea Basin and the intersection of these environmental problems with indigenous peoples of the Russian North contained in other chapters in this volume. The first part of the chapter provides a brief overview of human-environmental problems of the North. Next, northern areas with acute ecological problems are discussed. Then, information on northern air pollution, water pollution, and military, radioactive and chemical wastes is presented. In the process, pollution problems associated with forestry, the pulp and paper industries, mineral resource extraction and processing, the oil and gas industries, and defense industry waste products are noted. Other ecological problems generated from resource extraction and industrialization include disturbed vegetation, soil and ground; extensive regions of degraded habitat; and hence, significant real and potential threats to the biodiversity of the region. Some emission data and health statistics are used to assess the absolute and relative geographical severity of these pollution problems, as well as their possible impact on human health.

The chapter ends with a brief analysis of governmental and non-governmental institutions, laws and polices as well as some economic factors relevant to the creation of these serious problems. Environmental issues, challenges, and their likely trends are placed within the context of Russia's ongoing economic and social transition. Accordingly, such topics as the scale of long-term clean-up needed, the conflicts between profiting from the environment and protecting it, the potential benefits as well as costs of "market" reforms, the environmental threats of post-Soviet Russian development, and the potential and actual role of the international community are discussed.

General Aspects of Human-Environmental Problems in the Russian North

There are perhaps a half-dozen very important factors that characterize the human-environmental tensions in the Russian North. Perhaps most important is the legacy of Soviet economic development in the European North, Siberia and the Far East. Since the inception of Stalin's forced industrialization campaigns in the 1930s, these extensive, remote, resource-rich regions have been targeted for industrial development, mineral and energy resource extraction and processing which have had particularly disruptive and contaminating effects. This impact is also realized in surrounding ocean waters, as these regions drain primarily into the Arctic and Pacific Oceans Second, not only did Soviet development plans favor industrialization over traditional forms of economic activities, but all too often these industrial developments have been in conflict with traditional indigenous economic activities, such as reindeer herding, fishing, fur harvesting, and self-sufficient forms of agriculture, domestic animal husbandry, and logging, all of which require healthy ecosystems.

Third, the interplay of natural conditions in these high latitudes makes the Russian North very sensitive and vulnerable to many human economic disturbances. For example, much of the region has as long and severely cold winter seasons, extensive permafrost (*merzlota*) areas, limited precipitation, very modest evapotranspiration rates, and hence, very limited growing seasons. Accordingly, the predominant ecosystems indigenous throughout much of the Russian North are polar and alpine tundra, polar deserts, extensive swamps and marshlands, and extensive coniferous forest (taiga), much of it very slow-growing. Essentially all of these ecosystems may be classified as ecologically fragile. Once physically disturbed or contaminated, they are very slow to heal, regenerate and recover. Thus, resource extraction and processing such as mining, clear-cutting, pulp and paper manufacture, oil spills, ferrous and non-ferrous metal refining and the release of their various air- and water-borne wastes are even more harmful in these high latitudes than they would be in mid-latitude environments.

Fourth, since the devolution of the USSR foreign interest in resource extraction in Russia's northern regions has greatly intensified, which has proved to be a mixed blessing for Russia's environment. On the one hand, foreign technology is often less harmful and disruptive than former Soviet or current Russian resource extraction and industrial processing. On the other hand, the potential benefits of these more benign technologies are not certain to be realized for three major reasons. Firstly, some of the technologies allow for the economic extraction of even more remote and currently inaccessible resources, opening up relatively pristine environments to degradation. Secondly, reduced environmental costs per unit of resource output can be counterbalanced by the increased scale of extraction and processing activities. Thirdly, it is not always easy or cost-effective to adapt the foreign technology for use in the Russian economic landscape.

A fifth prominent factor is that new industrial facilities in the northern regions, instead of providing employment opportunities for local people, have relied on in-migration of industrial workers for the overwhelming share of their labor needs. Accordingly, as the negative impacts of northern industrial development have been felt, the opportunities for indigenous northern peoples to pursue their traditional economic activities and lifestyles has been continuously diminished and, as other chapters in this volume attest, this has led to the worsening of living conditions of local human populations.

Finally, although in the early 1980s all of the Soviet polar regions were rated as having "low" environmental stress impacts from industrial activities except for a "medium" rating for West Siberia (Gladkevich and Sumina 1981), unfortunately, as the following discussions suggest, the problems are now quite serious and there is little reason for optimism. Accordingly, unless these serious ecological-environmental problems in the Russian North can be successfully redressed, it will be impossible for Russia to honor international agreements it has already signed or to achieve sustainable economic development in the region, let alone provide more than a pretense of protecting the rights of aboriginal populations (Vilchek 1996: 21).

Northern Regions With Acute Ecological Problems

In the *Federal Report on the Status of the Environment in the Russian Federation in 1991* (*Gosudaarstvennyy doklad* 1992: 43-44) signed by Aleksandr Yablokov, Presidential Adviser of the Russian Federation for Policy in the Areas of Ecology and Health Protection, and Vitkor Danilov-Danil'yan, Minister of Ecology and Natural Resources of the Russian Federation, thirteen regions of the country are singled out as having acute ecological problems. Seven of these regions lie completely within the Arctic Ocean drainage. An eighth region, the "Middle Volga and Kama River" watershed, extends far to the north and is listed as having acute problems with depletion and pollution of surface and groundwater, disturbed earth by mining operations, soil erosion, ravine formation, atmospheric pollution, deforestation and degraded forest resources.

The ecological problems of the Kola Peninsula include disturbed earth by mining operations, depletion and pollution of surface and groundwater, atmospheric pollution, degraded forests and natural meadow lands, and violation of the regime for protected natural territories. The industrial zone of the Urals is cited for disturbed earth by mining operations, polluted air, depletion and pollution of surface and groundwater, contaminated soils, loss of productive land, and degraded forests. The industrial areas of West Siberia are indicated as having acute problems with lands disturbed by oil and gas extraction, contaminated soils, degraded reindeer pastures, exhaustion of fish resources and wild game, and infringement upon specially protected territories.

Farther south, the Kuzbass (essentially the Kemerovo Oblast' area shown on Map 5.1) suffers from disturbed land due to mining operations, atmospheric pollution, depletion and pollution of surface and groundwater, contaminated soils, loss of productive land, and soil deflation. Sadly, the famous Lake Baykal region makes this list because of both water and atmospheric pollution, depletion of fish resources, degradation of forests areas, gully and ravine formation, disturbance of permafrost soil layers, and infringement upon specially protected natural territories. The Noril'sk industrial district constitutes one of the most ecologically violated regions of the entire Russian North. The problems include disturbed land due to mining operations, pollution of the air and water, disturbance of permafrost soil layers, violations of protected forests, and deterioration of the nature-recreational quality of the landscape. Finally, the island of Novaya Zemlya makes this ignoble list as a result of radioactive contamination.

Air Pollution Problems and Prevention

Until recently, and unlike Western Europe and the United States, vehicle emissions were not a major source of air pollutants in Russia. While the emission gases of internal combustion engines are now becoming significant air pollutants in some large Russian cites, this is still not true for the Russian North, Siberia, and the Far East (with the notable exception of the city of Vladivostok). Nonetheless, more than 60 million people in the Russian Federation live in urban environments where the concentration of hazardous substances in the air exceeds the Russian Federation MACs (maximum allowable concentrations) by a factor of at least five; and of these citizens, 50 million live in cities with extreme air pollution—where the concentration of hazardous substances exceeds the MACs by a factor of ten or more. Only fifteen percent of Russia's citizens reside in areas where the ambient air quality meets the health standards (*Gosudarstvennyy doklad* 1992: 10).

In the vast northern areas of the Russian Federation, trans-regional air pollutants transported from Western Europe, southern Russia, Central Asia and Kazakhstan, China, Japan and Korea, as well as local forest fires, mining operations and industrial smokestacks are the culprits. Two of the quantitatively and ecologically most significant extra-regional airborne contaminants are nitrogen and sulfur oxides. These oxides react with atmospheric moisture and fall as acid rains. Table 5.1 shows the source regions and receiving regions in northern Russia for these problematic combustion gases on an annual tonnage basis during the early 1990s.

Table 5.1 Atmospheric Transport of Nitrogen (numerator) and Sulfur (denominator) Oxides (in 1,000 metric tons)

	Receiving region			
Source region	West Siberia	Krasnoyarsk Kray	East Siberia and Far East	Arctic as a whole
Europe minus European Russia	105/222	39/107	16/94	133/251
European Russia	94/144	295/53	15/48	21/115
West Siberia	136/300	56/60	27/48	15/25
Krasnoyarsk Kray	2/70	10/369	5/163	2/163
East Siberia & Far East	—/1	—/3	—/122	—/2
Kazakhstan & Central Asia	7/118	11/46	7/47	2/17
China, Japan, Korea	2/11	1/7	3/177	—/3

Source: Adapted from Vilchek et al. (1966), 29.

An easier way to interpret the ecological significance of these gases is given in Table 5.2, which shows the lands in northern or Arctic Russia disturbed by the fallout of chemical pollution. According to these data, acid precipitation impacts an equivalent of about 3.7 percent of the entire Russian Federation's territory. Vilchek et al. (1996: 28) claim that no less than 1.5 to 2 percent of the entire Arctic and adjacent regions are being subjected to *strong* environmental pollution fallout from extra-regional sources and that the extent of critical and acute areas is not less than 15 percent of the Russian Arctic region. They are not the only authors to voice the opinion that the scale of deleterious, anthropogenic pollution and other human activities threatens the ecological equilibrium of the entire region (e.g., Feshbach 1995; Wolfson 1994).

Table 5.2 Areas of Lands Disturbed as Result of Chemical Pollution of the Atmosphere in Russian Arctic and North (in 1000 km^2)

Region	Dust fall	Acid Precipitation
Kola Peninsula	20-30	100-120
East European North	10-15	90-100
West Siberia	3-3.5	4-4.5
Central Siberia	10.5-11	400-410
East Siberia	3-3.5	5-5.5
Northern part of Far East	2-2.5	2.5-3

Source: Vilchek et al. (1966), 30.

Recent discharges of harmful substances into the atmosphere from Russian sources located within the regions of the Arctic Ocean and parts of the Pacific Ocean drainage basins are listed in Table 5.3 and Table 5.4. As can be seen from Table 5.3 the Russian northern economic regions, namely, the North, Urals, West Siberia and East Siberia have atmospheric discharges per urban dweller (in kg/capita) ranging from 1.66 to 2.08 times above the Russian Federation overall average.

Table 5.3 Regional Discharge of Harmful Substances into the Atmosphere per Urban Resident (in kg/capita)

					Regional Rank (worst is 11, best is 1)			
Region	**1980**	**1985**	**1989**	**1992**	**'80**	**'85**	**'89**	**'92**
Russian Federation	427.2	375.4	328.4	NA				
North	792.3	708.3	684.1	NA	10	10	11	NA
East Siberia	806.2	825.0	681.8	NA	11	11	10	NA
West Siberia	626.0	624.8	600.9	NA	8	9	9	NA
Urals	763.4	623.8	545.5	NA	9	8	8	NA
Central Chernozem	391.0	342.5	309.6	NA	7	4	7	NA
Far East	356.5	317.2	300.5	NA	5	6	6	NA
Volga	374.4	267.0	215.6	NA	6	5	5	NA
Volga-Vyatka	232.2	204.3	187.7	NA	3	3	4	NA
North Caucasus	227.4	213.9	179.0	NA	2	4	3	NA
Central	235.4	183.9	137.9	NA	4	2	2	NA
Northwest	154.4	147.7	118.0	NA	1	1	1	NA

Source: Feshbach (1995), 18; and data from Ministry of Ecology and Natural Resources of the Russian Federation

Murmansk
Karelia
NW
Arkhangel'sk
NORTH
Komi Rep.
CENTRAL
VOLGA-VYATKA
Kirov
CC
Tatarstan
1
Perm'
URALS
Sverdlovsk
Tyumen'
WEST SIBERIA
VOLGA
Orenburg
2
3
Kurgan
NC
Omsk
Tomsk
Novosibirsk
4
Altay Kray
6
Kazakhstan
Uzbekistan
Turkmenistan
Kyrgyzstan
Tajikistan
Scale 1:12,000,000
0 100 200 300 400 500
Miles
0 100 200 300 400 500
Kilometers

MAP 5.1 Selected Economic Regions and Oblasts

A more detailed look at the relative severity of the impacts of atmospheric pollution on the urban population of northern Russia can be gleaned from looking at Table 5.4. This table contains data on the annual atmospheric discharges in kg/urban-inhabitant for twenty-eight regions, or all but two or three of the most southerly oblast's and republics which constituted the former Soviet economic regions of the North, Urals, West Siberia, East Siberia and the Far East. These data are extracted from data compiled for seventy-three Russian regions. Accordingly, any rank number of 37 or higher means that the region ranks in the worst half of all Russian Federation regions for a given year. Thus, twenty-four of the twenty-eight northern regions rank in the upper or worst half of all Russian Federation regions in terms of this crude measure of human impacts of air pollution, including eight of the ten worst Russian atmospheric discharge regions!

The recent reduction in the emission of air pollutants indicated on these two tables is associated almost exclusively with the "non-production" of industrial plants as a result of the many strikes, fuel and raw material supply shortages, and industrial plant closures caused by the breakup of the USSR, rather than any improved air-pollution abatement practices.

In addition to nitrogen and sulfur oxides, heavy metals (including copper, lead, mercury, strontium and others), and petroleum products, one of the worst and most widespread pollutants is the polyaromatic hydrocarbon, benzopyrene (BP), a known carcinogen. BP was detected in 160 of the 350 Russian Federation cities monitored by the Ministry of Health and the Ministry of Environment in 1991. The detected BP levels were at hazardous levels in ninety-two of these cities (*Gosudarstvennyy doklad* 1992: 6-13). Unlike the emissions of many other air contaminants which have declined since 1989 because of the economic downturn, the emissions of BP and dioxin have actually increased due to the low combustion temperatures being allowed in industrial and municipal thermal power plants. A very strong statistical link exists between urban air BP and dioxin concentration levels and malignant tumors (cancers) in Russian cities, particularly among the young inhabitants (Feshbach 1995: 3-9 to 3-12). In addition to thermal power plant emissions, the other main sectors responsible for air pollution in the northern regions of Russia include the chemical industry, ferrous and non-ferrous metallurgy, the fuel and energy sector, machine building, household heating, and transport. Specific economic sectors and cities in the Russian North, Siberia and the Far East experiencing significant health problems related to air pollution include: ferrous metallurgy in Magnitogorsk, Novokuznetsk, Nizhniy Tagil; the petrochemical and organic chemistry industry in Ufa and Sterlitomak; non-ferrous metallurgy (copper) in the Urals cities of Karabash, Kirovgrad, Krasnoural'sk, and Mednogorsk;

and non-ferrous metallurgy (copper-nickel) in the Kola Peninsula cities of Nikel' and Zapolyarnyy and Noril'sk in East Siberia.

Table 5.4 Regional Discharge of Harmful Substance into the Atmosphere per Urban Resident (in kg/capita)

					Regional Rank (worst=73, best=1)			
Region	1980	1985	1989	1992	'80	'85	'89	'92
Russian Federation	427.2	375.4	328.4	NA				
Krasnoyarsk Kray	1321.8	1423.3	1210.8	1273.1	71	72	72	73
Tyumen' Oblast'	1445.9	1471.6	1478.4	865.6	73	73	73	71
Komi Republic	1097.5	860.2	958.2	824.4	69	68	71	70
Chelyabinsk Oblast'	1037.1	949.2	821.4	613.7	68	69	68	68
Orenburg Oblast'	959.1	773.7	640.8	602.7	66	65	65	67
Murmansk Oblast'	850.0	859.9	649.2	582.6	64	67	66	66
Sverdlovsk Oblast'	768.2	690.6	666.4	500.1	60	63	67	65
Magadan Oblast'	770.7	600.2	550.5	448.6	61	60	63	64
Tomsk Oblast'	820.6	345.7	385.6	446.4	63	46	53	63
Arkhangel'sk Oblast'	388.7	371.6	446.4	439.7	44	49	57	62
Irkutsk Oblast'	617.1	560.8	423.7	388.0	57	57	55	60
Karelia Republic	407.2	450.7	466.3	375.7	46	54	58	59
Kemerovo Oblast'	673.0	655.7	527.7	367.8	59	62	62	58
Kurgan Oblast'	413.7	217.7	198.9	330.2	47	32	32	53
Perm' Oblast'	485.0	360.0	329.2	313.2	50	48	52	52
Bashkortostan Rep.	884.2	593.7	418.4	306.6	65	59	54	51
Omsk Oblast'	501.4	441.8	324.6	291.2	52	53	51	50
Chita Oblast'	170.9	181.7	172.6	259.1	20	25	24	46
Sakhalinsk Oblast'	384.0	461.5	427.7	248.7	43	56	56	45
Sakha Republic	331.6	298.3	294.4	244.6	38	40	46	44
Khabarovsk Kray	287.3	310.9	292.8	210.9	34	42	45	42
Buryat Republic	348.1	328.3	255.3	206.3	39	45	40	39
Primor'ye Kray	424.7	276.7	264.4	192.9	49	36	41	38
Tatarstan Republic	517.5	290.6	222.9	176.8	53	38	36	37
Kirov Oblast'	254.7	260.9	249.1	172.9	30	35	39	36
Amur Oblast'	168.1	220.2	218.7	159.7	19	34	35	34
Kamchatka Oblast'	177.2	181.2	174.9	156.4	22	24	25	32
Novosibirsk Oblast'	256.9	209.7	208.1	152.8	31	30	34	30

Source: Feshbach (1995), 18; and data from Ministry of Ecology and Natural Resources of the Russian Federation.

Information on northern cities with the highest levels of air pollution is listed in Table 5.5 by type of pollutant and source. For more detailed information on pollution in the Russian Arctic, especially the non-ferrous metallurgical center of Noril'sk—the city in *all* of Russia with the *most* contaminated air, the intense atmospheric pollution on the Kola Peninsula from non-ferrous metallurgy, in the Vorkuta area from the coal industry and cement production, in parts of the Komi Republic and northern Tyumen' Oblast' from oil and gas flaring, and in Arkhangel'sk Oblast' from the pulp and paper industry, see an exceedingly informative recent article in *Polar Geography* (Vilchek et al. 1996) and earlier works by Bond (1984) and Pryde (Bond et al. 1991: 403-412).

Table 5.5 Russian Federation Northern Cities Having the Highest Levels of Air Pollution by Substance, Industry and Level of Pollution, 1990*

Region/City	Substance	Pollution Source by Industry Branch
European North:		
Arkhangel'sk[b]	methy mercaptan, formaldehyde, SC_2	Pulp and paper
Kandalaksha	BP, NO_2, dust, SO_2	Non-ferrous metallurgy
Monchegorsk	formaldehyde, BP, SO_2	Non-ferrous metallurgy
Nikel' - Zapolyarnyy[a]	SO_2, formaldehyde, metals, NO_2	Non-ferrous metallurgy
Novodvinsk[b]	methy mercaptan, formaldehyde, SC_2	Pulp and paper
Severodvinsk[b]	methy mercaptan, formaldehyde, SC_2	Pulp and paper
Vorkuta[a]	dust, SO_2, NO_2, BP, hydrocarbons, metals	Coal industry, cement industry, chemical industry
Urals:		
Bereznikib	Hydrocarbons, SO_2, NO_2, NO	Chemical, mineral fertilizer
Chelyabinsk[a]	BP, formaldehyde, SO_2	Ferrous metallurgy, energy
Kamensk-Ural'skiy[b]	BP, hydrogen fluoride	Non-ferrous metallurgy
Kurgan[b]	BP, formaldehyde, dust	Petrochemical, metallurgy construction materials
Magnitogorsk[a]	BP, CS_2, dust, phenol	Ferrous metallurgy
Nizhniy Tagil[a]	BP, formaldehyde, phenol, ammonia	Ferrous metallurgy
Novotroitsk	BP, phenol, dust	(NA)
Orenburg	BP, dust, NO_2	(NA)
Perm'[a]	BP, formaldehyde, HF	Petrochemical
Sterlitamak[c]	BP, acetaldehyde, NO_2, H_2S, alphamethylsterol	Chemical, petrochemical
Ufa[d]	(NA)	Petrochemical, chemical
Yekaterinburg[b]	BP, formaldehyde, nitric oxide, NO_2, ammonia	Ferrous metallurgy, railway transport

(Table 5.5 continued)

West Siberia:		
Barnaul	BP, formaldehyde, dust	(NA)
Kemerovo[a]	BP, lead, formaldehyde, ammonia, NO_2, CS_2	Mineral fertilizer, chemical, ferrous metallurgy
Omsk[b]	BP, formaldehyde, ammonia, acetaldehyde	Petrochemical, chemical
Novokuznetsk[a]	BP, formaldehyde, dust	Ferrous, non-ferrous metallurgy, coal mining, energy
Novosibirsk[b]	BP, formaldehyde, NO_2	Auto transport, energy, construction materials
Prokop'yevsk[b]	BP, formaldehyde, dust	Coal mining
Surgut	formaldehyde, phenol, dust, NO_2, CO	Petrochemical
Tyumen'[b]	BP, formaldehyde, dust	Pulp & paper, construction materials, energy
East Siberia:		
Abakan[b]	BP, dust, formaldehyde	Heavy machine building, boilers
Angarsk[b]	BP, dust, formaldehyde	Medical-biological, auto transport
Bratsk[a]	BP, methyl mercaptan, CS_2	Non-ferrous metallurgy, pulp & paper, energy
Chita[b]	BP, formaldehyde, dust	Energy, chemical, machine building, boilers
Irkutsk[b]	BP, formaldehyde, NO_2	Energy, heavy machine building
Krasnoyarsk[a]	BP, formaldehyde, dust, NO_2	Chemical, non-ferrous metals, construction materials, auto-transport
Noril'sk[a]	SO_2, NO_2, Cu, Ni, Co, BP	Non-ferrous metallurgy, electricity production
Selenginsk[b]	BP, formaldehyde, hydrogen disulfide	Pulp & paper
Shelekhov[b]	BP, hydrogen fluoride	Aluminum refining
Ulan-Ude[b]	BP, formaldehyde, phenol, dust	Energy, construction materials, auto transport
Usolye- Sibirskoy[b]	BP, formaldehyde, NO_2, dust	Chemical, energy

(Table 5.5 continued)

Far East:		
Khabarovsk[b]	BP, formaldehyde, ammonia, NO_2	Energy, construction materials, petrochemicals, RR transport
Komsomol'sk[a]	BP, lead, formaldehyde, dust, phenol	Electricity, ferrous metals, energy, petrochemical
Yuzhno-Sakhalinsk[b]	BP, soot, NO_2	Energy, auto transport, boilers

*The Air Pollution Index (IZA) is an index determined as the sum of recorded annual concentrations of five substances in maximum permitted concentrations (MPCs) having the highest level, for each specific year, classified as dangerous and containing sulfur gas.

[a]Cities with consistently high levels of air pollution.

[b]Cities with Air Pollution Index of more than 15 (for 1989).

[c]Cities with more than five cases/year where PDK levels have been exceeded by factor of 10 for specific pollutants, or exceeded the PDK levels of three or more pollutants by a factor of 10 (IZA less than 15).

[d]Cities with an IZA less than 15, but with high levels of specific pollutants.

Source: USSR State Committee for the Protection of Nature (1989): 16-18; Sostoyaniye zagryazhneniya atmosfery na territorii SSSR v 1990 i tendentsiya yevo izmeneniya za pyatilentiye (1991): 118-123; Vilchek et al. (1996): 24-27.

In 1990, the Russian Republic government issued Resolution No. N93 which identified forty-three Russian cities where priority pollution abatement measures were desperately needed. The contamination levels of the two primary greenhouse gases, methane and carbon dioxide, were not even considered in this listing. Fully twenty-eight of the twenty-nine cities which have significant chemical industry production related to the defense sector are listed as having significant human health problems. Dioxin contamination is strongly suggested as being one of the chief causes of these medical problems. Finally, Russia's atmospheric releases of nine major industrial heavy metals and hydrogen fluoride (aluminum refining) at such places as Kandalaksha, Nadboitsy, Krasnotur'insk, Kamensk-Ural'skiy, Novokuznetsk, Krasnoyarsk, Bratsk, and Shelekhov as recently as 1992 ranged from over 750 percent to nearly 12,000 percent greater than combined EEC (European Economic Community) releases of these substances, respectively (ZumBrunnen 1997: 570).

In summary, three observations are worth noting. First, even excluding the Urals and Kuzbass industrial areas some of the very worst air pollution problems in Russia in terms of their impacts on both humans and the landscape exist in the Arctic region and adjacent to it. Second, the most positive improvement in urban air quality in the former Soviet Union and Russia was associated with the shift away from oil and coal towards natural gas for electricity generation and space-heating within major urban areas. Since the breakup of the USSR hard currency and domestic fuel shortages in some regions of Russia now threaten to create a significant backsliding in this regard. Third, while it is correct that recent foreign technological assistance has created the potential for modest improvements in ambient air quality, the preponderance of recent improvements represent only short-run pollutant reductions associated with the disruption of industrial production. However, the reduction in the arms race and, hence, the Russian armaments industry's reduced need for ferrous and non-ferrous metals could yield a significant environmental peace dividend by sharp and permanent curtailment of steel output from antiquated and dirty mills.

Water Resources—Pollution of Rivers and Seas

Although most of the major rivers in the Russian North are frozen over from two to seven or more months of the year and flow northward (an exception being the Kama River and its tributaries which drain the western slope of the highly industrialized Urals region) against the general east-west grain of human-economic transport, these rivers have long served as water transport routes. Because of the arduousness and cost of constructing and maintaining surface roads and railroads in much of permafrost-plagued Siberia, rivers were and are used during the short navigation season (three to four months) to transport much of the durable, bulky supplies to and products from settlements and cities located on the Ob'-Irtysh, Yenisey-Angara, Lena-Aldan river systems of Siberia. When thickly frozen, over these rivers also serve as "winter roads" to and from remote locations. In addition the Azov, Black, Caspian, Baltic and White seas are all connected for navigation as a result of nineteenth- and twentieth-century canal digging, channel dredging, and lock-and-dam construction.

For our purposes here, agricultural runoff, municipal and industrial waste discharges into water bodies are of far more critical concern than river transit. Over the past few decades the Soviet and post-Soviet press, technical journals, and Western scholars have published a large number of accounts dealing with specific water pollution problems.[1] These writings and the works upon which they are based make known that chronic water-quality problems exist in many lakes and rivers in Karelia and the Kola Peninsula, within many lakes and rivers in northern European Russia; many locations within the Ural-Volga-Kama-Caspian basin; along various rivers draining the heavily industrialized Ural Mountains; along the Ob'-Irtysh-Tom',

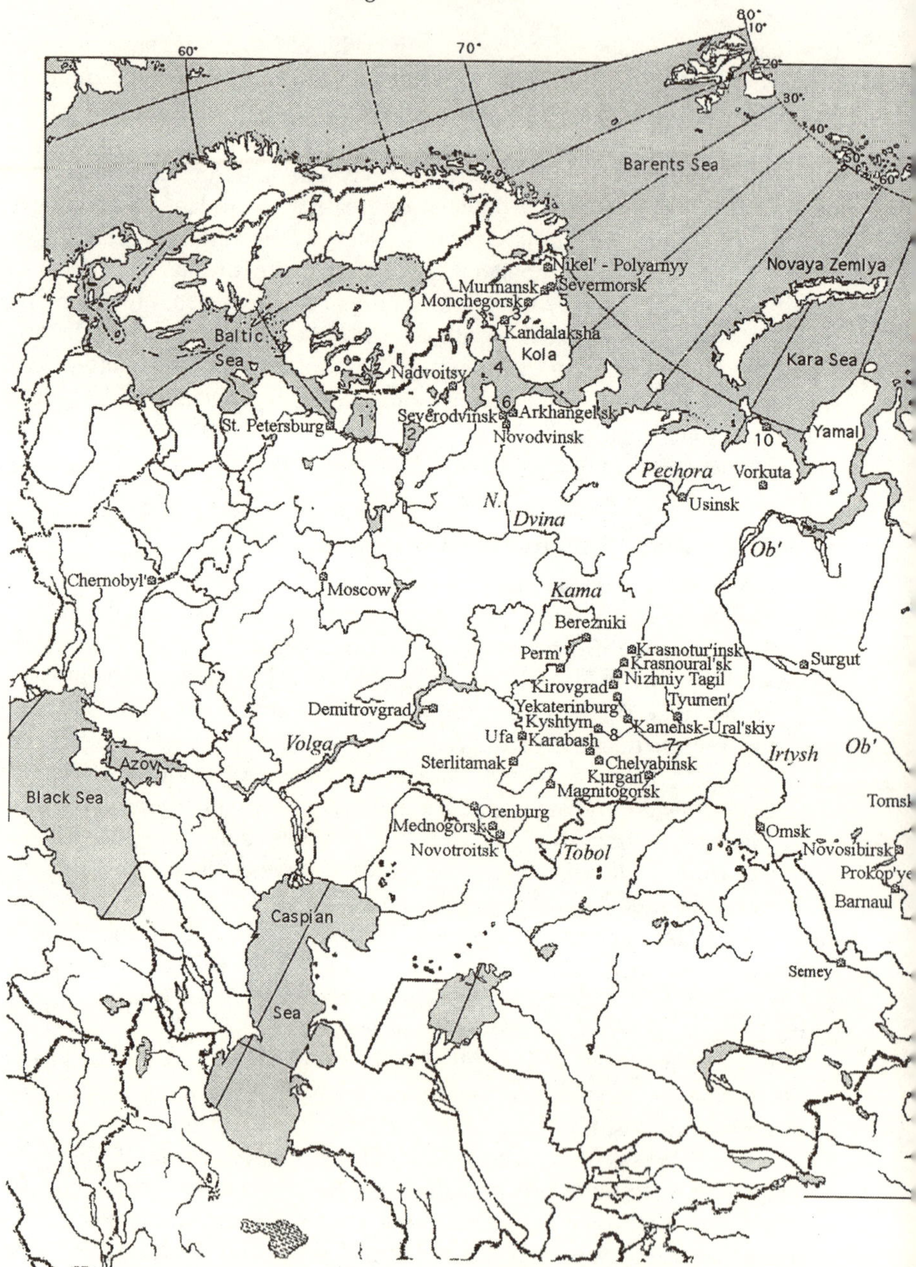

MAP 5.2 Cities and Rivers of the Russian North

Chukchi Sea
East Siberian Sea
Laptev Sea
Indigirka
Kolyma
Yana
Lena
Aldan
Lower Tunguska
Yenisey
Angara
Dikson
Noril'sk
Yakutsk
Ust'-Ilimsk
Bratsk
Krasnoyarsk
Abakan
Usol'ye-Sibirskoye
Angarsk
Irkutsk
Shelekhov
Baykal'sk
Selinginsk
Ulan-Ude
Chita
Komsomol'sk-na-Amure
Yuzhno-Sakhalinsk
Khabarovsk
Vladivostok
1 - Lake Ladoga
2 - Lake Onega
3 - Lake Imandra
4 - White Sea
5 - Teriberskiy Gulf
6 - Dvina Gulf
7 - Iset' River
8 - Techa River
9 - Tom' River
10 - Anderma
11 - Tiksi Bay
12 - Buor-Khaya Bay
13 - Yana Gulf
14 - Chaun Bay
15 - Pevek Bay
16- Lake Baykal
0
600
Miles

Yenisey-Angara river systems, other Siberian rivers and Arctic coastal areas, and the shores of Lake Baykal in southern East Siberia.

The most common pathogenic disinfectant used to deal with contaminated water is chlorine. Unfortunately, the chlorine can create potential dangers itself by reacting with any dissolved organic matter or compounds which may be present in the water to form such carcinogenic compounds as chloroform and trihalomethanes. In the Urals, in the vicinities of Ufa and Berezniki even dioxins and other toxic compounds reportedly have been formed in such ways in the surface wastewaters discharged from petrochemical plants. If true, these toxic compounds are likely to have been formed in many other surface water bodies as well. Elevated levels of organo-chlorines, such as PCBs, DDT, and HCH, and heavy metals, in some cases at toxic levels, are observed around the margins of the Arctic Sea, on the adjacent landmasses, and in the floral and faunal food chains in the region (*Gosudaarstvennyy doklad* 1992:14-20).

In 1991 approximately 73.2 cubic kilometers of wastewater were discharged into Russian Federation surface water bodies. Of this amount, 28 cubic kilometers of polluted water were discharged without any purification, 42.3 cubic kilometers had received only mechanical treatment, and only 2.9 cubic kilometers had undergone both mechanical and biological purification treatment prior to discharge. Of the wastewater effluents which pass through the woefully inadequate and inefficient municipal treatment plants, some 52 percent of the volume is composed of industrial waste streams, 34.5 percent of agriculture wastewater, and 13 percent of domestic sewage. About 50 percent of Russia's population drinks water that does not meet safety standards. One of the major problems compounding the problems of potable water quality is that both the sewer pipes and water supply pipes are in very poor condition in most Russian and NIS (Newly Independent States) cities. Accordingly, low water pressure is a chronic problem and contaminated sewage water enters the groundwater and thus enters the water supply network. In summary, in 1991 the USSR Ministry of Ecology acknowledged the following percentages of surface waters in the former Soviet Union exceeded the MAC (maximal allowable concentration) safety limits for various contaminants depending on location: ammonia—30-39 percent; copper compounds—70-75 percent; nitric acid—25-40 percent; oil and petroleum products—40-45 percent; organic substances—30-35 percent; phenols—45-60 percent; surface-active detergents—6-8 percent; and zinc—30-35 percent (*Gosudaarstvennyy doklad* 1992:14-17).

While these horrific data are aggregate statistics for the entire former Soviet Union, in relative terms, the sanitary conditions of very many surface water bodies in the northern regions of the Russian Federation are in relatively worse condition for two major reasons. First, something akin to an "out of sight, out of mind" phenomenon seems to have been operating in terms of the geographical distribution pattern of investment in pollution abatement and control facilities. In other words, except for some "high profile" examples such as the elaborate wastewater treatment facilities at the Bratsk (site visit by author, August 1983) and Baykal'sk pulp mills, investments in such environmental protection facilities long have had a lower priority at Northern and Siberian industrial complexes than in the more populated and highly developed European parts of the country (ZumBrunnen 1973). Second, the harsh winter conditions in much of the North and Siberia make biological treatment of sewage and industrial wastewater innately difficult and problematic.

More detailed information on pollution of the rivers and seas of the Arctic Basin can be evaluated making use of Tables 5.6 and 5.7. First, it should be noted that Russian MACs (PDKs in Russian) were/are based essentially on toxicological criteria geared to the fishing industry. As such they are quite lax compared with most international water quality standards (Levine 1961-66). Despite these loose standards the concentration levels for many contaminants in the rivers of the Arctic Basin watershed exceed the MACs (refer to Table 5.6), especially in the Indigirka and Kolyma (for phenol and ammonium), in the Yenisey (for oil, phenol, and zinc), in the Ob' (for oil, phenol, ammonium, and nitrogen) and even the remote Lena and Yana for phenol.

Table 5.6 Pollutant Concentrations (mg/liter) along Lower Reaches of Arctic Basin Rivers (averaged data for 1987-1993)

River	Ammonium ion $N(NH_4^+)$	Copper CU	Zinc Zn	Petroleum Products	Phenols
PDK (MAC)[a]	0.39	1.0	0.01	NA	0.001
PDK (MAC)[b]	2.90	0.005	0.005	0.05	0.001
Ob'	0.88	0.003	0.0025	0.35	0.002
Yenisey	0.43	0.005	0.03	0.40	0.003
Lena	0.04	0.003	0.005	0.05	0.004
Yana	0.03	0.003	—	0.07	0.004
Indigirka	0.60	0.004	—	0.06	0.04
Kolyma	0.68	0.005	—	0.025	0.003

[a]PDK = Maximum Allowable Concentrations (MAC) for surface waters.

[b]PDK = Maximum Allowable Concentrations (MAC) for sea waters.

Adapted from Vilchek et al. (1996): 31. Original sources: Shiklomanov & Skalalalskiy (1994):295-306; Izrael and Rovinskiy (1990): 111-112.

In addition to the river transport of pollutants into the various seas of the Arctic Ocean (refer to Table 5.7), atmospheric fallout and ocean currents, especially the North Atlantic Drift or Gulf Stream, contribute to the seas' problems. The still highly productive Barents Sea is the recipient of vast quantities of waste products and wood carried by the Gulf Stream from European dumping areas. While the open-water portions are still rather "clean," shipping lanes, several bays, and in particular the Kola, Teriberskiy and Motovskiy gulfs, are commonly contaminated (5-12 MAC) with petroleum products. A number of small rivers on the Kola Peninsula are highly polluted. For example, the average annual copper and nickel concentrations in the Nyudyay and Kolos-Yoki rivers exceed the MAC by 10-fold to 100-fold. Effluents from the chemical and pulp and paper industries are still contaminating Lake Imandra in the Kola Peninsula and Lakes Ladoga and Onega in Karelia. Lake Imandra is also heavily contaminated by copper and nickel (*Gosudarstvennyy doklad* 1992:16-17). Bottom sediment samples from the Barents Sea contain high levels of various harmful organic compounds (polychlorinated biphenols (PCBs) in range of 40-60 µg/g, pesticides up to 5 ng/g, petroleum products up to 3.5 mg/g, and phenols up to 5 mg/g) (Vilchek et al. 1996:30-31).

Table 5.7 Pollution of the Arctic Ocean by Major Contaminated River Discharges (in 1000s metric tons/year)

River	Nitrates	Phosphates	Total trace elements	Including Cu	Oil petroleum products
Ob'	25.40	18.00	78.89	5.50	162
Yenisey	5.62	5.88	89.19	3.00	232
Lena	225.00	3.57	80.65	1.50	60
Yana	1.52	0.254	3.90	0.29	—
Indigirka	1.912	0.559	7.07	0.25	—
Kolyma	4.96	0.782	8.10	0.34	—

Adapted from Vilchek et al. (1996): 30. Original source: Shiklomanov and Skalalalskiy (1994): 295-306.

The White Sea, or Beloye More, is reported to be still quite healthy despite significant wastewater discharges into it. The major problem seems to be in the Dvina Gulf which receives a number of wastes from the pulp and paper plants located along its shores (formaldehyde, methanol, tannin) and from municipal and other industrial sewage (copper, nitrogen compounds, phenols), and from ships (petroleum products). The Pel'shma and the Puksa rivers are both strongly polluted exceeding the respective MACs for ammonium nitrate, phenols, and ligno-sulfates (mine drainage from lignite leaching) by fifty to 100 times (*Gosudarstvennyy doklad* 1992:17)

In addition to the atmospheric discharges (polyaromatic hydrocarbons or PAHs, heavy metals, carbon and nitrogen oxides) from the flaring of up to 19 billion cubic meters of gas in West Siberia; multiple pipeline ruptures, transshipment spillage, and surface oil seepage have greatly increased in recent years. Environmentalists estimate that Russia annually may lose from 3 to 10 million metric tons of oil or in the range nearly one to three percent of its oil output annually through pipeline leaks into the ground and water bodies. An astounding number of pipeline ruptures occur, routinely upwards of 35,000 per year. Of these up to 300 are officially registered, meaning that each incident releases more than 10,000 metric tons of oil into the environment (Vilchek et al. 1966:26). The most problematic areas are the oil- and gas-producing districts in the Komi Republic in northeastern European Russia and in northern Tyumen' Oblast'.

The most well-known of the recent large oil spills occurred in August-September 1994 near the city of Usinsk in the northern part of the Komi Republic. While the official local government and Komineft (Komi Oil) Company claim only 14,000 metric tons leaked, the U.S. Energy Department claims 270,000 metric tons of oil gushed from this particular pipeline fracture. Internal documents from Komineft alone admit to over 1,900 oil leaks from faulty pipelines between 1986 and 1991. As a result, several tens of kilometers of the Kolva and Usa rivers and their tributaries' river banks have been covered in crude oil and the Pechora River has been threatened as well (Vilchek et al. 1966:26; Sagers 1994).

The Kara Sea basin has also been impacted by oil products and oil pipeline spills. The largest known recent catastrophe (500,000 metric tons of oil lost) happened in July 1990 near Belozersk in northern Tyumen'. Another pipeline break nearby at Nyagan' in 1993 spilled nearly as much oil (at least 420,000 metric tons). Nearly all rivers in the Ob' and Pur rivers basins of West Siberia are affected by oil pollution (ZumBrunnen 1997:567, Vilchek et al. 1966:26, Sagers 1994). In fact, petroleum products and phenols are officially reported to contaminate the entire length of the Ob' River from its source to its mouth. The average annual concentrations of petroleum products, ammonium nitrate, copper and zinc in the water of the Iset' River below the city of Yekaterinburg, Yeltsin's political home base, exceed their respective MACs by factors of ten or more. Farther east in the vicinity of the West Siberian city of Kemerovo, the Tom' River, one of the major tributaries of the upper Ob' basin, is recorded as being polluted by a number of highly toxic substances including aniline, varnishes, methanol and formaldehyde (*Gosudarstvennyy doklad* 1992:17).

The Yenisey River is the other major drainage basin from which polluted waters discharge into the Kara Sea. Two of the most problematic coastal areas are in the west in the vicinity of coastal port of Amderma and around the port of Dikson at the tip of the Yenisey Bay. Both are troubled by oil spills in the shipping lanes. Recently the measured levels of petroleum products, phenols, and industrial surfactants are reported to exceed their respective MACs by factors of 13, 10, and 7 in the Amderma area. In addition to petroleum products the estuaries of the Ob' and Yenisey rivers have been experiencing increased measured levels of heavy metals: copper, iron, lead, manganese, tin and zinc (Vilchek et al. 1966:31). Especially problematic for the Yenisey's waters below the city of Krasnoyarsk are lignosulfates (acid mine drainage) and volatile acids (*Gosudarstvennyy doklad* 1992:17). Despite large-scale capital investment in sewage-treatment facilities, the upper reaches of the Yenisey-Angara watershed have continuing problems with the effluents of pulp and paper mills, atmospheric fallout, and runoff from aluminum refining (the author visited some of these mills, refineries, and waste treatment facilities in the summer of 1983). Both the Ust'-Ilimsk and Bratsk reservoirs along the Angara are strongly polluted by concentrations of methylmercaptans and hydrogen sulfide (H_2S) 100-fold their MACs (*Gosudarstvennyy doklad* 1992:17). Finally, after more than thirty-five years of argument, scientific debate, and huge capital investment in pollution-abatement facilities, the pollution problems along the shores of Lake Baykal originating from factories at Baykal'sk, Selenginsk, and Ulan-Ude still remain far from benign (*Gosudarstvennyy doklad* 1992:17-18).[2]

Farther east in the Laptev Sea, the major pollution problems are along coastal bays and gulfs, with phenols released from immense quantities of submerged wood presumed to be the major source. Vilchek et al. (1996:32) list some extremely high measured phenol levels in a number of areas: 5 MAC levels at mouths of Lena and Yana rivers, 22 MAC levels in Bulunkan Gulf, 60 MAC levels in the Yana Gulf, 65 MAC in Buor-Khaya Bay. Contamination by petroleum products and untreated sewage are other problems. For example, petroleum products stain the waters of maritime shipping routes and exceed the MAC by a factor of 12 in Buor-Khaya Bay and by a factor of 20 in Bulunkan Gulf. Sanitary conditions in Bulunkan Gulf are claimed to be catastrophic. Tiksi Bay is heavily polluted by discharges of the settlements unpurified sewage (Vilchek et al. 1996:32). Recent official records indicate that the content of polluting substances in the upper course of the Lena River amount to from one to seven times the MAC and those in the middle and lower course from one to four times the MAC (*Gosudarstvennyy doklad* 1992:17).

In terms of both relative and absolute levels of pollution, the East Siberian Sea and Chukchi Sea are quite clean. Along the East Siberian Sea coastal margins, the only moderate problems are Pevek Bay and petroleum products concentrations equal to the MAC in Chaun Bay. Vilchek et al. (1996:32) report that the conditions of both the microbial and bottom fauna in Pevek Bay have improved over the past decade. During the annual spring high-water time, rains and meltwater flows in the Kolyma River have increased concentrations of harmful substances, especially manganese and lead (*Gosudarstvennyy doklad* 1992:17). The Chukchi Sea just north of the Bering Strait is still sufficiently "clean" to serve almost as a "pure ocean background" or "baseline" region. Nonetheless, the trends in measured pollutants has been upward in recent years. Despite their low concentrations in the sea water, even in these most pristine of the Russian Arctic coastal seas, there are concerns about the "biological magnification up the tropic levels" of several persistent organic chemicals, notably PCBs, DDT, HCCH, benzopyrene and other long-lived metabolites (Vilchek et al. 1996:32).

From a medical standpoint some of the most problematic contamination comes from river water in the Urals, the Northern Dvina, the Ob' and several of its tributaries, and the Angara. Water from these rivers have become documented sources of outbreaks of dysentery, cholera, viral hepatitis, typhoid fever, toxic and chemical poisonings (Feshbach and Friendly 1992:91-133, 181-203; Feshbach 1995). On the positive side, over the past two decades of Soviet power significant expansion of re-circulating industrial water supply systems and sewage treatment plant capacities occurred (*Gosudarstvennyy doklad* 1992:47-55). Ironically, perhaps the most positive improvements to water quality are associated indirectly with the precipitous downturn in the industrial production of Russia and the European NIS.

Military Radioactive and Chemical Wastes

Extremely important linkages exist between a number of Soviet and post-Soviet military activities and environmental problems. While international media headlines have flashed primarily upon the potential post-Soviet nuclear accidents, nuclear brain drain, arms exports, and potential nuclear terrorism (e.g., see Felgengauer 1996); similar concerns are warranted with regard to bacteriological and chemical warfare risks. The seriousness of the environmental contamination legacy of the Soviet military both in the former USSR and at former Soviet bases on foreign soil is reflected by the recent creation of an Ecology and Special Protection Systems Directorate under the Russian Ministry of Defense. As opposed to monitoring activities at test ranges, firing ranges, airfields, tank training areas, warehouses, and arsenals; these "green" helmets are charged with an active role in clean-up operations. Also included under this new

Directorate's jurisdiction are disaster relief operations, such as the post-Chernobyl' accident, earthquake relief, and ecological disaster relief operations (ZumBrunnen 1997: 572-573).

The Soviet nuclear industry, primarily run by the military, had a long, secret history of serious accidents before Chernobyl'. Safety had a low priority compared with weapons development and rapid commissioning nuclear power stations. Accordingly, nuclear power plants were built without radiation containment vessels (Dodd 1994). New information indicates that there have been many accidental atmospheric releases of radiation from the so-called Atomic Cities Network (*atomgrad*)—an estimated thirty-eight to eighty-seven secret cities initially established by Stalin to develop the atom bomb (Pryde 1991:36-55; Peterson 1993:140-150; Pryde and Bradley 1994; Rowland 1996).

In December 1975, while participating in a joint Soviet-American symposium on the use of mathematical models for water pollution control in Rostov-na-Donu, the author was approached by David Tolmazin, a Soviet physical oceanographer who subsequently emigrated to the United States, and told about a horrible "*tragediya*" (tragedy) in the Urals near Chelyabinsk. Tolmazin claimed that a huge area was completed fenced off and that there were people afflicted by radiation sickness who were living inside this enclosed space and not allowed to leave. He claimed many young people in their teens and twenties were missing their teeth and hair and looked decades older than their actual age. He said that the vegetation was dead over a vast area. He was so nervous and secretive in his speech and mannerisms and his story sounded so completely surreal and unbelievable to me that I was left totally perplexed. Four years later, however, along with the entire Western world I was to learn more about this *Nuclear Disaster in the Urals* when Zhores Medvedev (1979) exposed it to the West.

The tragic and "explosive" radioactive contamination history of the former USSR began not at Chernobyl' in Ukraine in April 1986, but rather decades earlier as Tolmazin had said "*ne daleko ot Chelyabinska*" ("not far from Chelyabinsk") at the Mayak Production Association (former names include Chelyabinsk-40, Chelyabinsk-65 and renamed Ozersk in 1994), near the city of Kyshtym in the Urals about ninety kilometers northwest of Chelyabinsk. This top-secret city constituted the core of the early Soviet atom bomb and nuclear weapons development program. Commencing operations in 1948, by 1951 76 million m^3 of liquid wastes containing 2.75 million curies of untreated high-, medium- and low-level radioactive processing wastes from the Mayak facility were directly released into the Techa River, a tributary of the Ob'. Radioactive concentrations successively downstream in the Iset', and Tobol' rivers were described as being 1/100 and 1/1000 that of the Techa, respectively. After traces of radioactivity were measured 1,600 kilometers downstream in the Arctic, the river was fenced off an some 7,500 people were evacuated from villages along the river (*Argumenty i fakty* 1989; Pryde and Bradley 1994:578-579).

Such direct discharges into the river ceased in 1951 and thereafter the radioactive wastes were diverted into the bottom of nearby Lake Karachay, a small, 260-hectare lake and several other artificial reservoirs. After about a decade the dumping of cesium- and strontium-laden nuclear wastes from the Mayak bomb-making factory resulted in this reservoir being completely filled containing (and leaking into the surrounding groundwater) 120 million curies worth of hazardous nuclear waste, an amount practically equal to 2.4 times the radioactive content of the debris released by the Chernobyl' accident. Thus, Mayak's engineers and directors embarked on a third waste-disposal method, making use of sixteen underground cement storage tanks. The first large-scale nuclear accident in Russia occurred at Kyshtym in September 1957 when a failure in the cooling mechanism for one of these nuclear wastes storage tanks resulted in a massive chemical steam explosion that ejected some seventy to eighty tons of radioactive material into the atmosphere containing about 2 million curies of the 20 million curies in the tank (Monroe 1992:533-538; Pryde and Bradley 1994:577-581; Peterson 1993:146-150).

The third major contamination event at Mayak occurred during the hot, dry summer of 1967 when Lake Karachay evaporated and winds dispersed the exposed radioactive dust with an activity of 0.6 million curies over an area of 2,700 km^2 and as far as 80 kilometers downwind, directly contaminating an estimated 41,500 people (Monroe 1992:538-539). Soviet sources admit that at least 130 million curies or 2.6 times the total amount of radioactivity released by the Chernobyl' accident have been released into the environment at Mayak (Pryde and Bradley 1994:578). As a result, approximately half a million people have been exposed to "elevated radiation doses," about 18,000 have been relocated, 935 have been diagnosed with chronic radiation illness and at least 217 of these victims have died (Monroe 1992:538). Medvedev (1991:113-114) postulates that tens of thousands of workers, possibly many of them prisoners, were involved in the clean-up of the Mayak-Kyshtym explosion site and so the total number of deaths may be far higher. As recently as 1990 radiation levels measured on the downwind lake shore, 600 roentgens per hour, were still sufficiently high to produce a lethal human dose within sixty minutes of exposure (Monroe 1992:538-539)!

Smaller scale accidents releasing plutonium into the atmosphere have occurred as recently as April of 1993 at Tomsk-7 in East Siberia and July of 1993 again at Chelyabinsk-65. Minatom (Russian Ministry of Atomic Power) has admitted that 32 to 33 million m^3 of wastes with a total radioactivity of 1 billion curies have been injected into the sandy Cretaceous strata at depths of roughly 310 to 340 meters and 240 to 290 meters from the Tomsk-7 facility. Furthermore, Russians have discharged up to 42,000 m^3/day of contaminated cooling water into the Tom' River. Ponds at Tomsk-7 contain 130 million curies, essentially the same as Lake Karachay at Mayak. Farther to the east along the Yenisey River near Krasnoyarsk lies Krasnoyarsk-26 (now renamed Zhelesnogorsk). Contaminated cooling water from the production reactors located there deep underground was discharged into the Yenisey River. As a result contamination levels are reported to be as high as 40 Ci/km^2 in the Yenisey below the facility and up to 3-5 Ci/km^2 along the lower reaches of the Yenisey (Monroe 1994:581-583).

Nuclear waste dumping into the waters of the Arctic north and the Pacific coastal seas, including the sea burial of spent naval reactors, combined with several nuclear submarine accidents, and the current storage of nuclear waste aboard ships anchored along coastal inlets in the northern and eastern seas are very serious problems indeed (Sagers 1991). Pryde and Bradley (1994:583-585) include both a table and map showing the locations of radioactive waste sites on Novaya Zemlya. Between 1964 and 1986 something like 11,000 containers holding "dangerous wastes" have been discarded into the Barents and Kara Seas. A whole series of nuclear fuels storage complexes are situated on the Kola Peninsula at Polyanyye Zori, and at the ports of Murmansk, Severomorsk, Litsa and on the island of Kildin (Pryde and Bradley 1994:586-587).

In addition to atmospheric testing of nuclear weapons, especially in the Semey (formerly, Semipalatinsk) region of Kazakhstan and on Novaya Zemlya in the north from 1949 to 1962, the Soviet Union set off 116 nuclear explosions for "technical purposes," such as dam construction projects, coal, gas and oil exploration, and mining projects between 1965 and 1988. For example, underground blasts were used to shatter rock formations in the diamond mines of Sakha (formerly Yakutiya). These blasts ranged from the equivalent of 1,000 tons of TNT (size of the nuclear bomb dropped on Hiroshima) up to 165,000 tons of TNT. Forty-three of these detonations were in the range of 5,500 to 11,000 tons of TNT. Despite being underground, several of these blasts heaved contaminated dirt into the atmosphere. Then, too, for over thirty years the Soviet Army conducted nuclear weapons and simulated nuclear explosion testing in the Heinamaa islands on Lake Ladoga, the source of St. Petersburg's drinking water (Pryde and Bradley 1994:583-589).

By far the most distressing example of massive Soviet and now Russian nuclear roulette is the recent disclosure of the secret injecting of billions of gallons of atomic waste, up to 3 billion curies or about half of the total accumulated nuclear wastes, directly into the earth at three sites. These sites are at Dimitrovgrad near the middle Volga, and at the previously noted Tomsk-7 facility near the Ob' River and the Krasnoyarsk-26 facility along the Yenisey River. The current radioactivity of these injected wastes is still estimated to be 1.45 billion curies. In the worst-case scenario, these radioactive wastes could leak to the surface and create large regional calamities downstream, contaminate large aquifers, and/or spread to the Caspian Sea in the case of the Dimitrovgrad site and to the Arctic Ocean in the latter two cases (Pryde and Bradley 1994:574-576). In the most benign and hopeful scenario, these wastes may stay deep underground for a sufficiently long time to render them more or less harmless by the long natural process of radioactive decay. Unfortunately, significantly elevated rates of cancer, especially leukemia, stillborns, and birth defects appear to be strongly associated with the populations living in or near these nuclear impacted regions and cities (Feshbach 1995).

The potential hazards from activities associated with chemical weapons are probably a greater human health threat than radioactivity, especially throughout Russia, Ukraine and the Baltic states. Such weapons were designed, tested, produced and stored in over 300 cities and towns of the former Soviet Union. Until at least the mid-1980s chemical weapons were dumped into the Sea of Japan, Sea of Okhotsk, the Black Sea of the Ukrainian coast, the Barents Sea, the White Sea, and the Baltic Sea. Soils, ground water and surfaces waters are widely contaminated by phenols, dioxins, and heavy metals associated with military weapons production. These contamination problems seem to be strongly associated with high rates of congenital abnormalities in many chemical-military related cities. Unfortunately, similar to other environmental problems, governmental funds available for the required clean-up and ameliorative efforts are grossly inadequate (Peterson 1993; Feshbach and Friendly 1992; Feshbach 1995).

The Environment and Social-Economic Transition

This brief overview of environmental pollution problems in the Russian North and Arctic drainage basins is very incomplete. Many environmental problems not discussed here, such as faunal and floral depletion and poaching, overfishing, overgrazing, destruction of spawning beds, lands disturbed by mining, land reclamation, transport development in permafrost zones, soil erosion, soil chemical contamination, timber harvesting and replanting rates, forest fire suppression, solid waste disposal and landfills, biodiversity, sustainable development issues, environmental legislation and regulation, land tenure legislation, property rights, eco-tourism, marketization and commodification of the environment, and a host of other resource management conflicts are indeed all major environmental problems which if anything have actually worsened since the dissolution of the Soviet Union (see Vilchek et al. 1996:32-41). There are number of interwoven factors which account for the aggravation of these environmental problems.

First, the centrifugal forces unleashed since the devolution of the Soviet Union have threatened the legitimacy of any and all laws, central governmental functions, agencies, regulations, and policies. Hence, enforcement of environmental laws and regulations has suffered nearly everywhere. In many instances what were at worst inter-Republic jurisdictional problems during the Soviet era have become international problems with few adequate dispute-resolution mechanisms in place. In the case of the regions discussed here—the Russian North and Siberia—their sheer vastness, the configuration of their river drainage basins and the prevailing wind patterns mean that the breakup did not generate many new international jurisdictional difficulties. On the other hand, the centrifugal forces unleashed by the USSR's breakup have engendered numerous and significant intra-Russia inter-regional resource ownership, resource control, resource management, and environmental protection conflicts (e.g., see: Kremlin sees its grip on regions slipping 1996; Centrifugal forces at work in Russia surveyed 1996; Why empowering regions is good: A Siberian view 1996; and Golubev 1996). Many of these conflicts have taken the form of the core (Moscow) versus the peripheries (the republics and oblast's). With the weakened central government and central command-economy institutions, major new inter-jurisdictional environmental management problems have arisen. Bond and Sagers (1992) provide a comprehensive analysis of the new Russian federation environmental protection law passed during the waning months of the USSR and signed by Boris Yeltsin on December 19, 1991, a mere six days before the Soviet Union was dissolved. Jurisdictional issues receive considerable attention in the new law, however, effective enforcement of the law and the consistent extraction of fines and others consequences for infractions are quite different matters that require political legitimacy across the landscape, financial resources and dedicated and trained personnel.

Second, even before the breakup of the Soviet Union, the financial resources necessary to solve or at least ameliorate these myriad problems were lacking, and the political will to do so was at best questionable. With the economic downturn continuing to worsen in many sectors of the economy and pessimism high (Survey finds mass pessimism about 1997 1997:1-3), the revenue collection systems are being severely "taxed" nearly everywhere (e.g., see Kulikov to lead crackdown on tax evaders 1997) and the social-political safety net is placing practically unprecedented demands on governmental coffers. Accordingly, the priority on redressing environmental problems has been diminished and the clean-up time table pushed far out into the future (Bisschop 1996).

Third, the various destatization and privatization processes may more accurately be called "personalization of wealth" processes. In other words, former Communist Party bosses, nomenclatura, and enterprises managers skillfully made use of their positions and connections to manipulate the privatization processes to become wealthy owners of former state enterprises, often leaving monopolistic entities intact (Frydman et al. 1996; Aslund 1996; Hansen 1996). Such windfalls and overnight highly skewed distributions of private wealth have created a new set of inefficient natural resource use problems (e.g., Natural monopolies 1997; Privatization II: No-holds-barred asset grab 1997).

Fourth, increasing globalization of the Russian economy and rising transportation costs have decimated a number of timber and other "non-commercially viable" resource extraction activities and enterprises in the North and Siberia (notable exceptions being oil and natural gas). This in practice had led to wanton disregard for a number of environmental laws and regulations and violations of protected environments as local citizens have been nearly forced or financially induced to engage in illegal and quasi-legal activities to eke out an existence. In other words, formerly officially protected natural, biological and environmental resources have become market commodities which can be and are being converted into hard currency. Examples, include poaching and effectively unregulated hunting, fishing, mining, timber harvesting, and all sorts of hard currency-induced resource smuggling. Low-pay and unpaid wages for various government officials and environmental regulators help make bribe-taking the order of the day.

Fifth, the transformation to a market economy naturally has created major "free rider" and other social cost problems in the form of private economic incentives to pollute the environment by casting off private production costs such as air and water-borne production wastes onto the environment or society as a whole. Whereas under the Soviet command economy the investment expenditures for pollution abatement equipment were borne by the state rather than an the individual (state) firm or enterprise; now such expenses really can affect a newly privatized firm's balance sheet. Even when such expenditures were granted by the state under the Soviet economy at no expense to an enterprise, enterprise managers still had a number of incentives to not install pollution abatement equipment in a timely manner (ZumBrunnen, 1974b; ZumBrunnen 1984). Without strong, enforceable pollution control laws and regulations, this situation is even worse for the environment under "free-market" conditions. Furthermore, many, indeed most, of the reform problems that DeBardeleben (1990) wrote about at the end of the Soviet period remain unsolved. On a positive note, in the highly profitable oil and natural gas industries of West Siberia, less polluting Western technologies are being put into practice as foreign loans often require such infrastructure investments.

Sixth, the lack of clearly specified and codified property rights in land, a stable and fair tax system, debt financing, and effective securities and bankruptcy laws hamper both efficient and equitable environmental management decision-making (e.g., Kulikov to lead crackdown on tax evaders 1997; Bekker 1997; Pismennaya 1997).

Finally, there are a number of Russian domestic environmental education, perception, scientific, and technological issues which thus far have become more acute since the devolution of the USSR. These negative trends may take a decade or more to reverse under the most optimistic scenarios.

Conclusions

Environmental pollution has played a major role in the virtually irreversible degradation of up to two percent of the land area of the Russian North or Arctic. The most severely impacted areas are in northeastern European Russia, northern West Siberia, the southern Taymyr (Noril'sk region), industrial cites in the Urals, industrial cities in the Kuzbass, cities along the upper Angara valley, and on the Kola Peninsula. Vilchek et al. (1996) argue that another 25-30 percent of the Russian Arctic is in a critical situation due to human-induced pollution, degradation of near-tundra forests, losses of biological diversity, and exhaustion of some biological resources. In terms of the Russian Arctic basin's freshwater resources, the most polluted and overfished are the Northern Dvina, the Pechora, the Ob' and the Yenisey. The Barents Sea's marine ecosystem is the one most stressed due to commercial biological harvesting. All of the Arctic seas remain relatively clean except for several bays, gulfs and coastal zones contaminated by human and industrial pollutants.

The majority of the unsolved environmental problems that loom in the Russian Arctic's future can be linked to the numerous interrelated "institutional" factors listed in the previous section. However, the physical and biological dimensions of these unresolved problems include the inadequate monitoring of the new oil and gas developments in West Siberia and the Yamal Peninsula, mineral raw material developments in several locations, potential large-scale radioactive contamination of aquifers and surface freshwater, timber-harvesting techniques in sensitive near-tundra forests, the massive need for oil-spill clean-ups, installation of modern air and water pollution control facilities and incentive structures for their effective operation, environmentally sensitive and appropriate transportation modes and networks, and most importantly, that elusive and undefinable concept of a unified sustainable development scheme. While the Russian North and Siberia suffered grievously under communism, the covetous hands of global capitalists are equally capable of raping Siberia of her riches (Linden 1995)! While several USAID- and EEC-funded projects are underway to try to come to grips with the tragic environmental legacy of the Soviet era and the new environmental management problems and issues engendered by the economic changes underway in Russia and the NIS, their combined financial and personnel resources are far from being equal to the task. One of the harshest criticisms of these various foreign assistance programs is that the vast majority of the funds allocated has ended up in the pockets of Western consultants.

Finally, while many may argue that the weakened Russian institutions will eventually allow the freedom needed for Adam Smith's efficient "hidden hands" to modernize and transform the economy and democratize Russia simultaneously, this author still has strong concerns about the pervasive new environmental "free riders" and the rampant "overt feet" of organized crime or "the mafia" strangling, bleeding, and kicking the environment to near death (e.g., Kulikov takes on corruption in the aluminum industry 1997).

---

[1] For example, see: DeBardeleben (1985), Feshbach and Friendly (1992), Goldman (1972), Komarov (1980), Peterson (1993), Pryde (1991, 1995), Wolfson (1994), Ziegler (1987), and ZumBrunnen (1973, 1974a, 1974b, 1978a, 1978b, 1984, 1987a, 1987b).

[2] For an early account of the Lake Baykal controversy, see: ZumBrunnen (1974a).

References

Argumenty i fakty. 1989 (34): 8.

Aslund, Anders. 1996. Reform vs. 'rent-seeking' in Russia's economic transformation. *Transition: Events and Issues in the Former Soviet Union and East-Central and Southeastern Europe* 2 (2): 12-16.

Bekker, Aleksandr. 1997. State to accept shares as payment of debts. Defaulters will have a choice. *Current Digest of the Post-Soviet Press* 49 (10): 15.

Bisschop, Gita. 1996. Optimism wand for a prompt cleanup. *Transition: Events and Issues in the Former Soviet Union and East-Central and Southeastern Europe* 2 (10): 42-45.

Bond, Andrew R. 1984. Air pollution in Noril'sk: A Soviet worst case? *Soviet Geography* 25(9):665-680.

Bond, Andrew R., Matthew J. Sagers. 1992. Some observations on the Russian Federation Environmental Protection Law. *Post-Soviet Geography* 33(7):463-474.

Bond, Andrew R., Mark Bassin, Michael J. Bradshaw, George J. Demko, Leslie Dienes, Paul Goble, Gary Hausladen, Ronald D. Liebowitz, Philip R. Pryde, Lee Schwartz, Victor H. Winston. Panel on Siberia: Economic and territorial issues. *Soviet Geography* 32(6): 363-432.

Centrifugal forces at work in Russia surveyed. 1996. *Current Digest of the Post-Soviet Press* 48 (48): 5-6.

DeBardeleben, Joan. 1985. *The environment and Marxism-Leninism.* Boulder: Westview Press.

———. 1990. Economic reform and environmental protection in the USSR. *Soviet Geography* 31(4):237-256.

Dodd, Charles. 1994. *Industrial Decision-Making and High-Risk Technology: Siting Nuclear Power Facilities in the USSR.* Lanham, MD: Rowman and Littlefield.

Felgengauer, Pavel. 1996. A lucrative spot: The Russian system of weapons trading. – Only middlemen get rich. *The Current Digest of the Post-Soviet Press* 48(38):1-4.

Feshbach, Murray, ed. 1995. *Environmental and health atlas of Russia.* Moscow: "Paims" Publishing House. Map 2.31: Integrated evaluation of ecosystem stability (ecosystem vulnerability).

Feshbach, Murray and Alfred Friendly, Jr. 1991. *Ecocide in the USSR: Health and nature under siege.* New York: BasicBooks.

Frydman, Roman, Kenneth Murphy and Andrzej Rapaczynski. 1996. Capitalism with a comrade's face. *Transition: Events and Issues in the Former Soviet Union and East-Central and Southeastern Europe* 2(2):5-6.

Gladevich, G. I. and T. I. Sumina. 1982. Measuring the environmental impact of industrial centers in the natural-economic regions of the USSR. *Soviet Geography: Review & Translation* 22(3):155-163.

Goldman, Marshall. 1972. *The spoils of progress: Environmental pollution in the Soviet Union.* Cambridge, MA: MIT Press.

Golubev, Vladimir. 1996. How are you getting along, provinces?: 'Counteroffensive' against Moscow. Regions answer the VCHk's demand for payment of taxes by issuing an ultimatum. *Current Digest of the Post-Soviet Press* 48(48):7.

Gosudarstvennyy doklad: O sostoyanii okruzhayushchey prirodnoy sredy Rossiyskoy Federatsii v 1991 godu. 1992. [*Federal report on the status of the environment in the Russian Federation in 1991.*] Moscow.

Hansen, Philip. 1996. Structural change in the Russian economy. *Transition: Events and Issues in the Former Soviet Union and East-Central and Southeastern Europe* (26 January) 2(2):18-21.

Izrael, Yu. A. and F. Ya. Rovinskiy, eds. 1990. *Obzor sostoyaniya okruzhayushchey prirodnoy sredy v SSSR* [*Survey of the condition of the natural environment in the USSR*]. Moscow: Gidrometeoizdat.

Komarov, Boris, pseudonym of Ze'ev Wolfson. 1980. *The destruction of nature in the Soviet Union.* White Plains, NY: M. E. Sharpe.

Kremlin sees its grip on regions slipping. 1996. *Current Digest of the Post-Soviet Press* 48(48):1-4.

Kulikov to lead crackdown on tax evaders. 1997. *Current Digest of the Post-Soviet Press* 49(7):1-5.

Kulikov takes on corruption in the aluminum industry. 1997. *Current Digest of the Post-Soviet Press* 49(8):7-8.

Levine, B. S., ed. and trans. 1961-66. *USSR literature on water supply and pollution control.* 9 vols. Washington, D.C.: U.S. Public Health Service.

Linden, Eugene. 1995. The rape of Siberia. *Time* 146(10):42-53.

Medvedev, Zhores. 1979. *Nuclear disaster in the Urals.* New York: W. W. Norton.

Medvedev, Zhores. 1991. Do i posle tragedii [Before and after the tragedy]. *Ural* 4:97-116.

Monroe, Scott D. 1992. Chelyabinsk: The evolution of disaster. *Post-Soviet Geography* 33(8):533-545.

Natural monopolies: Can they be tamed? 1997. *Current Digest of the Post-Soviet Press* 49(7):5-9.

Privatization II: No-holds-barred asset grab. 1997. *Current Digest of the Post-Soviet Press* 49(8):5-7.

Peterson, D. J. 1993. *Troubled lands: The legacy of Soviet environmental destruction.* Boulder: Westview Press.

Pismennaya, Yevgenia. 1997. Government is pleased with good financial results. *Current Digest of the Post-Soviet Press* 49(10):15-16.

Pryde, Philip R. 1991. *Environmental management in the Soviet Union.* Cambridge: Cambridge University Press.

———. ed. 1995. *Environmental resources and constraints in the former Soviet republics.* Boulder, Westview Press.

Pryde, Philip R and Don J. Bradley. 1994. The geography of radioactive contamination in the former USSR. *Post-Soviet Geography* 35(10):557-593.

Rowland, Richard H. 1996. Russia's secret cities. *Post-Soviet Geography and Economics* 37(7):426-462.

Sagers, Matthew J. 1991. Nuclear waste illegally dumped at sea off Novaya Zemlya. *Soviet Geography* 32(10):706-707.

———. 1994. Oil spill in Russian Arctic. *Polar Geography and Geology.* 18(2):95-102.

Shiklomanov, I. A. and B. G. Skalalalskiy. 1994. Studying water, sediment and contaminant runoff of Siberian rivers. In *Workshop on Arctic contamination May 2-7,1993*. Anchorage, Alaska: Arctic Research of the United States (8):295-306.

Sostoyaniye zagryazhneniya atmosfery na territorii SSSR v 1990 i tendentsiya yevo izmeneniya za pyatiletiye [The state of pollution of the atmosphere upon the territory of the USSR in 1990 and the tendency of its change over the five-year plan]. *Meteorologiya i gidrologiya* [Meteorology and Hydrology] 4:118-123.

Survey finds mass pessimism about 1997. 1997. *Current Digest of the Post-Soviet Press* 49(3):1-3.

Vilchek, G. E., T. M. Krasovskaya, A. V. Tsyban and V. V. Chelyukanov. 1996. The environment in the Russian Arctic: Status report. *Polar Geography.* 20(1):20-43.

Why empowering regions is good: A Siberian view. 1996. *Current Digest of the Post-Soviet Press* (25 December) 48(48):6-7.

Wolfson, Ze'ev. 1994. *The geography of survival: Ecology in the post-Soviet era.* London and Armonk, New York: M. E. Sharpe.

Ziegler, Charles. 1987. *Environmental policy in the USSR.* Amherst: University of Massachusetts Press.

ZumBrunnen, Craig. 1973. *The geography of water pollution of the Soviet* Union. Ph.D. dissertation, U.C. Berkeley.

———. 1974a. The Lake Baikal controversy: A pollution threat or a turning point in Soviet environmental consciousness? In *Environmental deterioration in the Soviet Union and Eastern Europe,* ed. Ivan Volgyes. New York: Praeger Publishers.

———. 1974b. Institutional reasons for Soviet water pollution problems. *Proceedings of the Association of American Geographers.* 6 (April).

———. 1978a. Geographic-economic factors in water pollution control systems. In *American-Soviet symposium on use of mathematical models to optimize water quality management.* Gulf Breeze, Florida: Environmental Research Laboratory, Office of Research and Development, U.S. Environmental Protection Agency.

———. 1978b. An estimate of the impact of recent Soviet industrial and urban growth upon surface water quality. In *Soviet resource management and the environment,* ed. W. A. Douglas Jackson. Columbus, Ohio: AAASS Press.

———. 1984. A review of Soviet water quality management: Theory and practice. In *Geographical studies on the Soviet Union: Essays in honor of Chauncy D. Harris,* eds. Roland Fuchs and George Demko. Chicago: The University of Chicago, Dept. of Geography, Research Paper No. 211.

———. 1987a. Soviet water, air, and nature preservation problems of the Gorbachev era and beyond. *The Soviet economy: A new course?: NATO colloquium, Brussels,* ed. Reiner Weichhardt. Brussels: NATO.

———. 1987b. Gorbachev, economics and the environment. In *Gorbachev's economic plans, volume 2.* Washington, D.C.: U.S. Government Printing Office.

———. 1997. Russia and the European NIS. In *Contemporary Europe: A geographic analysis*, 7th ed., ed. William Berentsen. New York: John Wiley & Sons.

6

Negotiating Nature in Swedish Lapland: Ecology and Economics of Saami Reindeer Management

Hugh Beach

This paper concerns the political dimensions of the ongoing debate about overgrazing and range deterioration in Swedish Lapland.[1] Reindeer herding in Sweden is a livelihood reserved exclusively for those of Saami ancestry. As recognized in Swedish legislation, reindeer-herding rights are all that remain in practice of Saami indigenous rights. Therefore, recent environmentalist claims of destruction by reindeer of the Swedish mountain habitat and new state regulations devised to protect ranges against what is regarded as Saami mismanagement, strike at the heart of Saami claims, cultural maintenance and self-determination. Many Saami have come to regard the ecology promoted by the state as yet another instrument of Swedish colonialism.

While scientists often regard the so-called problem of the commons from the perspective of an objectified, given nature that must be sustained by social regulations, my meaning is that such a perspective obscures the political formation of nature itself. The supposedly given environment is itself a matter of negotiation beyond the domain of *its* exploitation or preservation.[2]

Material in this paper has largely been gathered from Saami reindeer herders in the field, continuously from 1973-77, and intermittently thereafter. Research over the past three years in particular, focused upon overgrazing issues, has been made possible by funding from the Nordic Environmental Research Program and the Joint Committee of the Nordic Social Science Research Councils.

When pressed for content for their ideal "natural norm," state authorities refer facilely to the latest ecological slogans: sustainability and biological diversity. Yet, at the same time, state herding authorities continue to apply strong measures for the so-called rationalization of the reindeer industry to promote sustainable, but maximal, economic yields. Maximal yields will require that sustainability be stressed to its extreme; and even should maximal economic yields be harnessed at the breaking point of sustainability, this condition is certainly not conducive to the promotion of greatest biodiversity.

Newly instituted fines for herd-size transgressions as well as numerous forms of subsidies to the herders have been implemented in the effort to patch the equation between ecology and economy. However, the fines, as structured in the Reindeer Act of 1971 and its subsequent revisions, cannot inhibit potentially destructive herder competition for scarce grazing. And the subsidies carry the onus of depicting the Saami herding livelihood as an economically pointless hobby maintained by the Swedish taxpayers.

Given this scenario, the moral justification for a herding livelihood at all would come to rest solely on the rights of minority cultural preservation as upheld in international law. At the same time, Swedish rationalization policies and the free march of modernization propel herding toward full-blown ranching and technologies similar to that of other Swedish animal farming, developments which to the majority population have little or nothing to do with Saami culture and therefore should not imbue herding with any special indigenous resource rights. While much of the Saami counteroffensive against perceived Swedish "ecolonialism" confronts the state on its own terms, e.g., with claims that Saami herd management is in fact environmentally sound, Saami are with good reason vehement against being enfolded within a purely Swedish ecological framework.

Ecology or "Ecolonialism"?

As a result of the UN conference on environment and development in 1992, *Agenda 21* was adopted which urged all nations to produce strategies for sustainable development. The participation of indigenous peoples in the formulation of such national environmental strategies was emphasized. The Swedish Ministry of Environmental Protection received directives from the government in 1995 to present measures to attain sustainability in the country's mountain regions. In the resulting report entitled "Sustainable development in the country's mountain regions" (SOU 1995:100), the overly simplistic concept of sustainable development[3] is applied to Swedish Lapland and focuses to a considerable degree explicitly on reindeer herding.

> It is important to understand that a sustainable development encompasses *ecological and social, as well as, economic* development. Those who live in the mountain regions must be given the possibility of work, education, security and quality of life *within the limits that nature can tolerate.* (SOU 1995:100, p. 8; my translation and my italics)

While this position seems eminently reasonable, it takes for granted that an unregulated population, "those who live in the mountain regions," with links to the market economy can live comfortably and sustainably by utilizing nature's surpluses without in so doing affecting, changing, or actually defining nature. In fact, as the very existence of the report illustrates, a trade-off is demanded between the long-term preservation of nature and the short-term welfare of its human constituents. The report (SOU 1995:100) has resulted in a government bill to Parliament (Prop. 1995/96:226) of the same name and with basically the same recommendations which at this time (Sept. 1996) awaits Parliamentary process.

In the early summer of 1995 a large earthslide occurred on Stora Axhögen mountain in the Funäsdalen valley within the territory of Mittådalens Sameby. The prominent Swedish environmentalist Nils G. Lundh attributed the earthslide to overgrazing by reindeer. Soon thereafter, in a debate article, Margareta Ihse, a natural geographer, made the following statement based on her study (begun in 1993 for the World Wildlife Foundation) of vegetation changes in certain herding districts:

> During the last decades, the exploitation of the mountain regions has been discussed with growing intensity. The increased tourism with recreational building, alpine centers, trails and snowmobile traffic has occasioned increased pressure on a sensitive mountain nature. At the same time, that reindeer herding which has been practiced for thousands of years has altered character. It has experienced great changes. A traditional herding undertaken by nomadic Saami has been replaced by modern management and rationalization with mainly settled Saami and greatly increased numbers of reindeer as a result. In combination this increases pressure on the sensitive mountain nature. . . .
>
> . . .The WWF report warned that such rapid erosion might also come to pass in the Swedish mountains. The newly observed earthslides show that this is the case and that it is urgent that solutions be found. (Ihse, in *Samefolket* 1995:14)

In the latter half of the 1800s, with the demise of the so-called "parallel theory," whereby it was thought that Saami herding and Swedish settlement could exist side-by-side without conflict, a massive body of legislation was enacted to minimize and to mediate the conflicts between herders and farmers. Detailed regulations were constructed to control herd movements and herder responsibilities for the protection of crops. As Swedish farming in the North ebbed away under the "rationalization" policies following

World War II, the basis of regulating herding on that score declined (Beach 1980).

The next regulatory justification was founded on the growing socialistic welfare ideology. Medical studies in the 1950s had shown that the "vital statistics" of the Swedish Saami were comparable to those of underdeveloped nations. Saami infant mortality, for example, pulled down the national average, and Sweden sought to solve the problem by devising a comprehensive program of rationalization for herding's structure and production embodied in the Reindeer Act of 1971 (often abbreviated RNL) (SFS 1971:437, updated as SFS 1993:36). Regulations were now oriented toward increasing the living standard of Saami herders—a praiseworthy goal, but to a great extent one pursued at the expense of diminishing the herder population (fewer mouths to share the same limited resources) (Beach 1983).

Today we are at the threshold of a new regulatory framework of the herding livelihood, that based upon environmental concern. The Reindeer Act has been embellished by a new paragraph (§65a) with so-called ecological objectives. Other new sections have been added to the Reindeer Act in the attempt to further control herd numbers and to decrease the risks of overgrazing (Proposition 1992/93:32). Statements by environmentalists and Swedish politicians make it clear that they regard proper ecological theory and practice as something belonging to the domain of Western science. If the Saami are at all recognized as practicing ecologists, they are considered to be poor ones.

> Sustainability demands, however, that reindeer herding take consideration of environmental conditions. It is both of national and international interest that the natural resources be utilized in a balanced and controlled manner. The limits for reindeer grazing should be coordinated and adapted according to the reindeer's natural wanderings, so that the ecological goal can be achieved. The *partial* responsibility of reindeer herding for this process is clear. Too much pressure on the grazing lands can lead to ecological damage and contribute to the destruction of the basis for reindeer herding. Therefore the educational base on environmental issues should be broadened among the Saami. (Sápmi, the Social Democratic Party's political program regarding the Saami, 1995)

Added to the understandable environmentalist motivation that the protection of nature is the business of all people, and not just that of populations local or indigenous to the area in question, is the motivation provided by international law. What Eide terms a "maximalist" interpretation of Article 27 of the Convention on Civil and Political Rights holds states responsible for protecting the resource base which is the precondition for the exercise of the rights of indigenous peoples (Eide,

1985:204). Of course, the Saami welcome the support of international law when it comes to protecting the reindeer grazing lands against massive destruction, for example, by the timber industry, but they now face a paternalistic twist whereby the same forces of protection are invoked against their own livelihood for their own good!

Newspaper headlines such as, "Trample of reindeer herds destroys the mountains," (*Norrbottens Kuriren*, Dec. 9, 1994) or reports that overgrazing by too many reindeer is transforming the mountains into a pile of rocks (*Dagens Nyheter*, July 21, 1995) have become commonplace. Herders are blamed for the decimation of scarce species (reindeer predators) and the use of high-tech equipment (such as motorbikes) in herding that destroys the tundra. Overgrazing and range destruction by the trampling of too many deer (or the motorized vehicles to herd them) have become major concerns of environmentalists, not just herding administrators who have had a long involvement with the problem on the grounds of protecting the sustainability of the herding industry. The issue has now grown far beyond this; concern does not stop with the viability of herding but with the fate of the Swedish mountains. As before, under the Swedish welfare ideology, the new phase of regulations is advertised to the Saami as being for their own good. Now as then, it is a truth with qualifications.

A number of hot issues dominate relations between the Saami and the Swedish regulators. They can usefully be broken down under two categories, the one concerning arguments over the equality or exclusivity of resource access, and the other having to do with environmental degradation. In the former category are items such as tourism, hunting and fishing. (New hunting regulations, not expropriating, but confiscating exclusive Saami hunting and fishing rights in the regions above the Agriculture Line were recently put in place.) In the latter category we have the debate over use of high-tech equipment among herders, the right to life of reindeer predators, tourism again, and the overgrazing issue. All issues involve discussion about both principles and practice. For example, should the Saami be permitted to develop their herding "industry" as they see fit as can so many other businesses? Or should they be forced to use only traditional methods in keeping with the principle of cultural preservation, the basis on which the state has granted them special resource privileges (Beach 1993a)?

The way in which the Saami manage or mismanage the resources to which they enjoy special rights of access impinges directly on the sympathies the Saami might cull among the majority voting population and, by extension, on the bills enacted by Parliament concerning matters of resource rights and access. For the Saami, a minority, and for the herding Saami, a minority within a minority (and there are indeed major conflicts between herding Saami and non-herding Saami) moral arguments are essential. It is one thing to have a legal right, but quite another to keep it. Arguments used by the Saami to counter the equality-of-access supporters involve symbolic rhetoric such as accounts of national parks trashed by swarms of tourists. They point out the destruction of wildlife due to the new open-hunting regulations (Sametinget 1994), and they can invoke pro-indigenous international conventions ratified by Sweden (Beach 1994).

Those opposed to Saami interests can invoke arguments and symbols of the Saami as ecological "fallen angels." They point out that current Saami resource utilization is far from traditional and hardly different from what any Swede might do if given the chance. They confront in the courts Saami land ownership claims and claims of immemorial rights of usage, as for example in the ongoing Härjedalen conflict (Beach 1985, 1992; Cf. Sveg Case lower court verdict 1996).

Of course there are many sincere environmentalists who, rightly or wrongly, view certain aspects of Saami livelihoods as destructive of a natural habitat, the fate of which, regardless of any ownership debates, affects us all and is rightly a common concern. Naturally, given this situation, there are also those who find it convenient to affect an environmentalist stance, not out of concern for the habitat, but in order to discredit Saami moral arguments, to undermine special Saami rights of resource access with that justification, and ultimately to exploit these limited resources yet further. Just as naturally, a given Saami stratagem can be to discredit environmental arguments by insinuating ulterior motives. Unfortunately, even sincere environmentalists have sometimes felt their scientific integrity violated by accusations that they are anti-Saami racists. Among those points lobbied are: Which species and livelihoods are to be considered "natural?" To which forms of exploitation might one turn a blind eye, and for how long, before demanding sacrifices on the part of these or other forms of livelihood for the common ecological good? And for whom is nature anyway?—a question whose very formulation harbors dubious metaphysical presuppositions.

Placing "Eco-" in Economy: the Problem of the Commons

A number of terms must be defined in order to proceed. With reference to the goal ranges of a system (the ranges of states in which systems remain stable) mentioned by Rappaport, or its "limits of flexibility" according to Bateson, we can speak of "goal ranges" for the reindeer-grazing ecosystem, one inclusive of grazing and reindeer (*not herders*). When related to the reindeer element of the system, one can speak of "goal reindeer ranges" to signify the lowest to the highest number of reindeer which can be sustainably accommodated in the reindeer-grazing ecosystem.[4]

Another term related to herd size and used frequently by the herding authorities is "rational herd size." In Sweden, the land on which reindeer can be grazed is divided into about fifty so-called Samebys, which besides defining territorial units, also define social and to some extent economic herding units. Rational herd size is the term commonly used to define for a Sameby the greatest number of reindeer (of an age/sex composition to yield the greatest yearly profit) which can be regularly sustained on the seasonal range (usually the winter range) that forms the bottleneck in the Sameby's annual grazing cycle—that is, without endangering regenerative capacity of the pasturage. What is most rational at any time, of course, varies according to the integrated cost/profit shifts of numerous variables (see pp. 15 ff.). "Rational herd size" is a term whose origin lies squarely with the herding authorities, is grounded in the concepts of the Western market economy, and does not concern itself with satisfactions other than dollars and cents (within sustainable bounds). For an in-depth historical review and critique of rational herding policy, see Beach (1981). While an accurate rational herd size fits within a Sameby's goal reindeer range, it is not necessarily identical to its upper limit. Goal reindeer ranges take no consideration of the age, sex, size or meat quality of the reindeer utilizing the land. A Sameby sustained at its maximal goal reindeer range would probably in turn sustain its herders to a lesser degree than it would were the herd kept at the rational herd size.

Finally, there is the "total allowable reindeer quota" or (TAQ), the figure set by the herding authorities as the ceiling herd size permitted for a Sameby. TAQs are tailored to each Sameby according to the best educated guess of its bottleneck seasonal grazing capacity. This includes supposedly hard data from grazing inventories, but also past experience. Like the rational herd size (which unfortunately is often considered to be one and the same with the TAQ even in principle), a TAQ is supposed to be within the goal reindeer range of a Sameby, pressing its upper limit. While ideally the TAQ and the rational herd size of a Sameby are to coincide, TAQ values are quite stable and do not follow the rapid shifts of the market, slanting maximal profits toward the production of fewer, but bigger deer or increased numbers of smaller head. While a TAQ might be adjusted should a Sameby suffer major and permanent loss of territory due to the building of a hydroelectric dam, for example, TAQs are generally the same year after year, their margins of error recognized to be greater than most adjustments which might be justified by market shifts.

Given the fact that each of Sweden's fifty-odd Samebys has an individually designated total allowable reindeer quota (TAQ),[5] instituted precisely to obviate overgrazing, it is plain that the private herd size of any one herding Sameby member, his herding labor engagement, husbandry

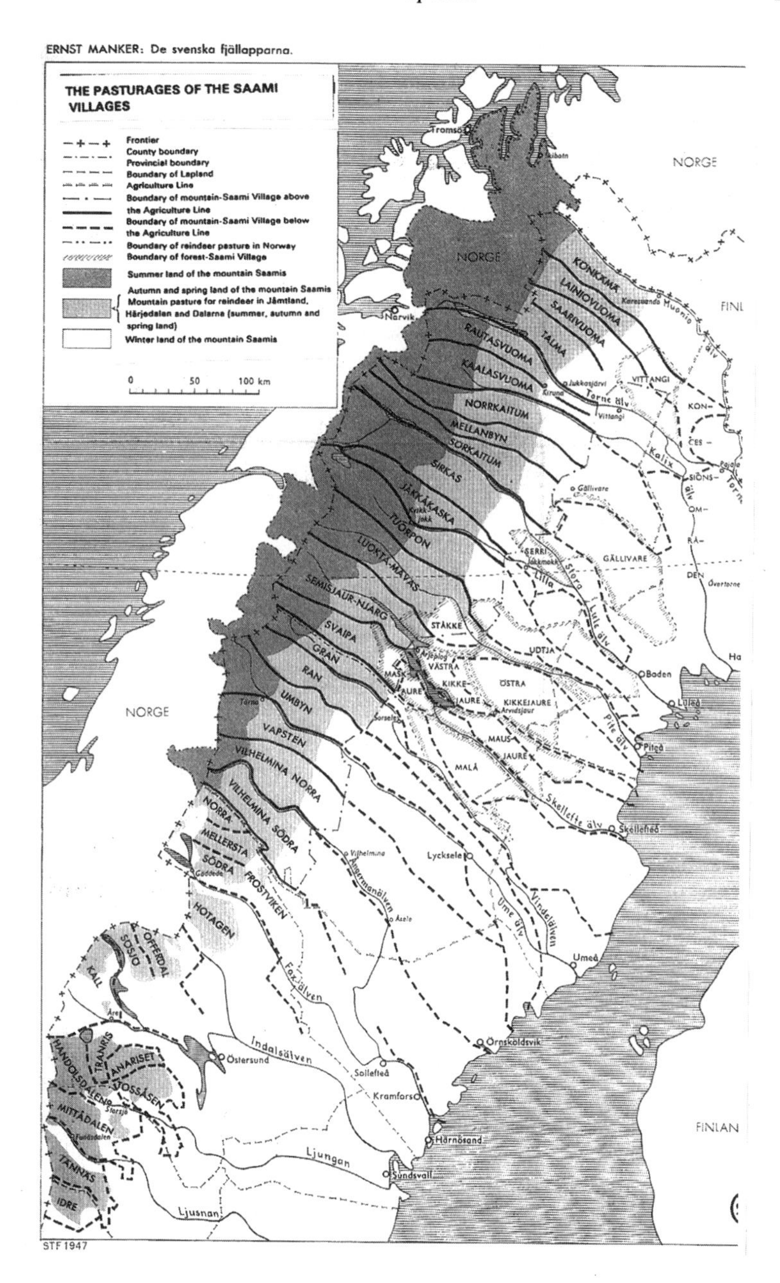

MAP 6.1 Samebys

goals and success at realizing them, affect directly issues of herd size and management forms for all his other Sameby fellows.

Here as in other pastoral societies, a basic goal of the pastoralist is to maximize his herd size both for reasons of prestige and security. The risk of reindeer losses to predators or to the "bad winters" mentioned above motivates many herders, who own reindeer stock privately but who graze them on lands held in common by the Sameby group, to opt for an expansionist ideology. Moreover, the Swedish law regulating the herding livelihood, the Reindeer Act of 1971 (with subsequent revisions) bases a herder's voting power within the Sameby upon his herd size (justified by the same arguments that give a shareholder power in a company proportional to his or her share of stock in it), but with the stipulation that herders get one vote for each newly started hundred head of deer.

These conditions fit nicely the preconditions of the Hardinian model of commons dilemma leading to eventual tragedy of the commons. Environmentalists, especially their opportunistic, uncritical supporters, have not been lax in pointing this out. Yet, it does not follow that all cases of reindeer grazing tragedy are caused by the commons dilemma. Furthermore, biased opinions of northern farmers who prefer not to have reindeer in their fields, or of hunters who do not want reindeer interfering with their hunt, or of the timberland owners who do not want to negotiate with herders over their herding rights before beginning logging operations easily transform their claims that there are "too many reindeer" for their own self-interests into claims that there are too many head for the grazing lands. News of winter reindeer starvation, whether caused by climatic grazing blockage or not (see below), serves further to confirm the impression of tragedy of the commons.

Of course, this is not to say that such claims are categorically wrong or that the self-interests of those opposed to the herding livelihood and Saami special resource rights might not overlap with environmental interests and a correct assessment of tragedy of the commons. Claims of tragedy of the commons must be given careful consideration. Reindeer in excess of the Sameby TAQ provide only a short-hand, crude indicator, but one that is nonetheless legally compelling. Herders are known to tend to fudge the accuracy of their required annual herd counts (often for purposes of avoiding tax), and even with the best of intentions accuracy might be low under so-called "extensive" management forms (see below). Also, TAQs hold margins of error in relation to maximal goal reindeer ranges, and, as we have seen, such simple quantifiable values without considerations of variable herder pastoral mediation cannot be adequate measures for assessing real grazing tragedy.

The problem of the commons has attracted attention and been revisited a number of times, not least in anthropological fora, even before its succinct logical formulation by Hardin. It continues to fascinate, because in it we perceive not simply a puzzle whose solution(s) would entail practical economic benefit, but also in a nutshell the dilemma of purposiveness and rationality of living systems as members of a dynamic ecological whole. As such, the problem of the commons is one that confronts the evolutionary process with one of its most difficult challenges: how can a social category of behavior evolve (for example, to regulate the commons) which on one contextual level negates the lower-order optimizing behaviors of its individual constituents?[6]

The Problem of Defining the "Natural Norm"

We must proceed cautiously in this analysis when using such concepts as overpopulation and overgrazing, for these terms might not only be falsely proclaimed, but they often conflate important distinctions. For example, while it is commonly assumed that when reindeer starve to death it is due to an overly high reindeer/grazing ratio, starvation often occurs simply because ample grazing has been rendered temporarily inaccessible by crusted snow or ice lumps. The situation is all the more complex, since the aspects of grazing presence and grazing availability are tightly integrated. Not only are lichens (in Sweden generally the most limited grazing resource and therefore the bottleneck to herd increase) spread unevenly over the winter grazing lands, but climatic conditions frequently vary so much that while availability might be blocked in one area, it is undisturbed in another. Obviously, should a rise in reindeer numbers cause a strain on the overall reindeer/grazing ratio, the effects of "bad winters," where grazing availability is locked in broad patches, will become increasingly severe and will appear to be more and more frequent.

The herder mediates the relation between his reindeer and the grazing lands (Paine, 1994). Reindeer grazing preferences vary seasonally. Certain types of grazing follow a seasonally variable pattern of availability. In any one year a herder must regulate where and to what extent his herd utilizes a specific area of pasturage. He may well have to economize in one area at one time so as to have some left there at a crucial juncture later on. Moreover, a herder must be aware of grazing possibilities and alternatives throughout his range, not only for one year, but for many years. He must take all of these factors into consideration along with the economic needs of his family, the movements of other herders and the grazing pattern of their herds, seasonally available transportation routes and means, seasonal physiological changes of his animals, and numerous other factors. There are as many different herding strategies as there are herders, in fact far more, since a single herder will avail himself of many strategies and must be able to switch on the spot.[7]

Because of the profound effect of the herder's mediation between his herd and the grazing lands, it is impossible in simple linear fashion to relate herd size and grazing pressure or depletion. Herders exercise different degrees of so-called intensivity or extensivity over their animals, sometimes keeping them gathered under tight control (intensive herding), at other times allowing them to disperse and mix at will over a wide region only to be brought together for such things as calf-marking, herd separation, or slaughter (extensive herding). A number of animals held together intensively will exert a different pressure on the grazing lands (both in terms of grazing and trampling) than the same number dispersed under an extensive herding regime. Nor is it an easy matter to relate intensive or extensive degree to grazing effect, for in the long run, extensivity (though *prima facie* less wearing on the environment) might result in erratic husbandry practices (for example, opportunistic slaughtering rather than slaughtering based on careful selection) and even unnecessary herd increase. I have previously termed "over extensivity" (Beach 1981) the situation whereby herding skills are lost and a downward spiral of decreasing profits and increasing expenses leads to less time available for herding and therefore further extensivity. Or, the same extensive spiral with its increased demands for the use of costly "high-tech" equipment in herding operations can force the herder who is determined to earn his main income from this livelihood to increase his herd size in order to increase his profits simply to maintain the same standard of living for his family.

The point I wish to emphasize is that a broad spectrum of relatively sustainable herd-management forms with different herd sizes or similar herd sizes, but with different intensive/extensive methods, is possible for the same area. Similarly, a herd held in a more or less sustainable relation with its grazing land under one management form might well be propelled into an inherently unstable relation which slowly depletes grazing resources by another management form despite uniform reindeer numbers. Quantifiable values of reindeer numbers and calories available on grazing areas, taken alone, are but hopelessly crude indicators of grazing pressure.

Average reindeer slaughter weight and individual reindeer health are prime indicators of grazing depletion. Overly high grazing pressure will make itself apparent in the condition of the deer. Such effects are indeed undeniable, as are the existence of other homeostatic mechanisms which take effect with grazing depletion, such as increased predation, and lower birth and calf survival rates (Ingold, 1976:30-32). However, we are once again confronted with multiple systems of sustainability. The same grazing resource can host sustainably a maximal herd of well grown, fat animals, or maybe a herd of twice that number of more scrawny, stunted animals (with pastoral mediation as a further accommodating variable). Herders are forever discussing the current and historical sizes of the deer in different regions, sizes which have been known to vary considerably, just as we might note that the average European today is larger than his counterpart was in the Middle Ages.

Again, my point is simply that while extreme malnutrition can of course prove fatal to reindeer, there is a broad zone of healthy, acceptable (and hence economically and politically defined) reindeer physiognomies. What kind of deer comprises the standard for the TAQs? The period regulated by the principles of structure and production rationalization mentioned above based all considerations of herd size, herd age and sex composition as well as slaughter decisions (e.g., calf-slaughter, because calves yield more meat per grazing consumed than do full-grown deer) squarely on principles of maximum market profitability within sustainable ecological constraints. A large proportion of male calves especially were to be slaughtered prior to their consumption of scarce winter lichens and when, as with all reindeer, growth abates with the approach of winter. The herders were given incentives by way of subsidies to follow what was then seen as the course of maximal market profitability. For example, in the early 1970s a set subsidy was established for payment to the herder for each deer slaughtered regardless of size. In this way the pay per kilogram brought in through slaughter of a calf was enhanced in comparison to that brought in through slaughter of an older, heavier deer with the same subsidy. Since the institution of this subsidy, the realities of the market have altered drastically, whereas the subsidy form has not. Reindeer meat is now classed by quality, and the slaughterhouses have an intricate differential price scale according to class and export potential. Calf meat, which cannot meet the proportion of meat to bone favored by restaurants, is now classed way down. Subsidies and rationalization programs, however, still try to steer herders in the direction of calf slaughter. I can only imagine that the policy continues on the grounds that despite the reduced appreciation of calf meat, it is thought that herders might still make more money from maximal winter herds predominately composed of pregnant, calf-producing females than from winter herds containing, for example, also many mature bucks and castrates.

Note that this argument might be valid in a Sameby operating at its maximal TAQ—not counting those calves slaughtered as they do not live to burden the most common situation of winter lichen bottleneck—but not at all necessarily valid for a Sameby with total herd size well below that. A Sameby which operates below its TAQ, and which need not economize with its grazing, will likely profit more by letting its calves live on through the winters and grow to maturity before slaughter. Their greater weight will bring more profit, and their meat quality will bring a better price per kilogram probably regardless of the calves' subsidy enhancement. Grazing conservation in this instance is not a problem, and the growth intensity of calves no longer a factor to consider (Beach 1981). In short, the reindeer is not a God-given standard unit, but one which to great extent is composed and promoted not only by environmental conditions, climate, and the mediation of the pastoralists, but also through market fluctuations, pricing schemes and subsidy policies. This is a situation common for traditional livestock, many of which have been bred and fed into extreme human creations. Although not so exaggerated, similar considerations are still at work with reindeer, even if they roam freely much of their lives, and even if the policies which control their physiognomy are quite variable in the degree to which their consequences match their conscious intent.

Gradually, but with increasing momentum, the reindeer is being transformed into a farm-like animal. Road networks penetrating far west into the mountains, along with the development of mobile slaughter trucks, have revolutionized slaughter procedure and castration policy. Now, to accommodate the market, this is followed by the increased use of artificial fodder, not only as a catastrophe measure in the face of bad winters, but also as a regularly scheduled pre-slaughter fattening measure.

Regular use of artificial winter fodder is also seen by the promoters of rational production as a means to widen the winter grazing bottleneck caused by limited lichen availability to match the greater grazing capacity of other seasons and thereby to make significant gains in TAQs. Not only does the use of fodder tend to eliminate the homeostatic effects of bad winters on herd size, but with a possible rise of winter herd size, it will increase the dangers of overgrazing dramatically should foddering for some reason be discontinued or impossible to realize. Fodder not only makes larger herds possible, it also requires increased production—more reindeer—to pay for it.

Although reindeer herding in Sweden has long been directed toward and dependent upon a market economy (while still preserving elements of the subsistence livelihood with regard to the herding family's own food production), the degree to which the reindeer as biological individual is impacted by market demands has been minimized by herder traditions and the costs involved in exerting the necessary control over at best only semi-domesticated livestock well able to thrive in wilderness regions with little or no human contact. Market-induced changes have been far more readily observable in matters such as age/sex composition of the total herd. A well-functioning system of rational calf slaughter implies a winter herd of maximally explosive reproduction in the spring, one that is dominated by calf-bearing females with only sufficient mature males to ensure their impregnation. Naturally, removal of the older males from the herd affects both the needs of the herd but also its fund of experiential knowledge. This in turn will have considerable bearing upon the performance of management, and all of these factors will impact the energy demands of the herd—labor and financial demands on the herders but also demands on the grazing resource. Herds deprived of the experience of older male reindeer might well lose more in cost-efficiency than they gain by conserving the grazing they would otherwise have eaten for growth-intensive calves. In short, the grazing pressure of reindeer on the land will vary depending upon management form, physiognomy of individual animals, and composition of the herd. In establishing TAQs and promoting herding rationalization standards for management within such limits, one cannot avoid setting reference values for these variables and hence setting what is to be considered the reference value for the natural state of the grazing lands, for these are formed by (not simply reduced or conserved by) such policies.

In an article commenting upon that of Margareta Ihse, Anders Sirén states:

> Mountains without reindeer are equally unnatural as mountains with too many reindeer. The Fulu Mountain's thick lichen carpet hinders seeds from other plants from growing and is a product of the extermination of the wild reindeer, a key component of the ecosystem, through hunting. (Sirén, in *Samefolket* 1995:16)

Infuriated by what they regard as an attack on their livelihood and a threat to their ethnic minority rights, Saami have countered on a number of points the assumption that the Axhög Mountain earthslide was occasioned by reindeer overgrazing. They point out that similar earthslides have occurred previously in many places where no reindeer graze at all (Andersson, in *Samefolket* 1995:13), and that even if reindeer have grazed in the area, there are other factors, for example extreme weather conditions with exceptional spring flooding, of far greater significance in causing the earthslide (Andersson, in *Samefolket* 1995:12). Others, such as the Saami herder Johansson (quoted by Andersson, in *Samefolket* 1995:12f) and the non-Saami researcher in ecology and environmental protection Sirén (in *Samefolket* 1995) both admit that erosion caused by the trampling and overgrazing of reindeer certainly does occur where the reindeer are confined in large numbers at particular sites during corrallings. However, they both object to the gross generalization of attributing these limited, local problems to the Swedish mountains as a whole. Most interesting in this context, however, is another point made by the herder Johansson:

> Yes, of course, lichens disappear in the summer mountain grazing areas where the reindeer are, he admits. But this here is not something which is an evil for

> reindeer herding. The lichens are replaced by green plants which the reindeer eat during the summer time and are of course much better. The comparisons made between the mountain regions of Norway and Sweden are like comparing a cow pen with an uncut meadow. (Johansson cited by Andersson, in *Samefolket* 1995:12)

Johansson's point is that at least some degree of trampling and grazing of lichens does not necessarily induce erosion in the summer lands; in fact the green plant replacements have a much more developed root system which binds the earth better than lichens. Moreover, green-plant replacement may be more advantageous for the total grazing system of the Sameby should this rather than lichens be the bottleneck factor. Of course where lichens are trampled and/or overgrazed so might green plants be. Johansson does not shy away from admitting the problem in principle. He questions whether it has been given the proper emphasis.

With regard to the point raised here, Johansson's comment is illuminating, for it recognizes that the reindeer not only consume nature, they form it. If, as has been noted, thick lichen cover in summer mountain pastures hinders other vegetation from taking hold, then obviously "overgrazing" of these lichens might mean exposed surfaces and erosion, at first. But other green vegetation might thereby have the opportunity to seed and in time create a more erosion-resistant cover. Yet it would be wrong to characterize the environmentalist block as being innocent of the concept of a quasi-domesticated nature (i.e. one that is conditioned by humans as well as the animals in their service). The Swedish Environmental Protection Committee's report fully recognizes not only that the reindeer contribute to the formation of the natural landscape, but also that this formation is not necessarily one of reducing its riches by eating it or trampling it:

> it is probable that there is to a certain degree a positive correlation between grazing pressure and biological diversity, that is to say, a completely ungrazed area contains fewer species than a moderately grazed area, while a further increase in grazing pressure leads to a reduction of diversity. (SOU 1995:100, p. 13)

The Saami journalist Andersson sums up the obvious rejoinder:

> Certainly one must admit that the reindeer leaves its mark on nature, but this it must be permitted to do if one is to be able to carry on with reindeer herding. Then one can ask oneself what a natural condition can be. Given an area which has been utilized for reindeer grazing for many hundreds of years then surely it must be considered a natural state. (Andersson, in Samefolket 1995:13)

Nor would it be correct to characterize the environmentalist block as not accepting the principle that a quasi-domesticated nature can be worthy of protection. In her article debating the overgrazing issue, Margareta Ihse states:

> The preservation of the nature of the mountains is not just a question concerning the Saami. The Swedish mountain regions are important and unique in a European perspective. Now, during the European nature preservation year, Sweden will receive international recognition for her nature-preserving actions in protecting and caring for meadows and enclosed pastures in the agricultural landscape. (Ihse, in *Samefolket* 1995:15)

Again the matter reverts to negotiated emphasis—not simply concerning how much human-oriented purposive behavior is to be allowed to steer nature, but also as to when such determinism is still to be accepted and preserved as part of the natural norm. A considerably larger degree of human determinism is likely to be protected within the framework of natural preservation in the so-called agricultural landscape than would be tolerated as creating the natural norm in the mountains.

Herding for Biological Diversity

According to a memorandum from the Board of Agriculture addressing the matter of regulating reindeer numbers, some effort, if vague, has been made in a government bill (Prop. 1994/95:100 Bil. 10) to define what should be considered the natural norm. The issue is at least acknowledged (even if one comes hardly closer to a specification of what should be the natural norm by referring to that which is "representative" and should be preserved to "sufficient" extent).

> When the highest allowed reindeer number is decided, the point of departure shall be what nature can sustainably bear so that the *biological variety can be preserved* and a persistent utilization of the grazing resource not be forfeited...so that the vegetation is not impoverished and so that *representative types of nature are preserved to sufficient extent.* (Constenius & Danell, 1995:1, my italics)

We might term the herd size recommended above, that exerting moderate grazing pressure and conducive to biological diversity of pasturage, as the "diversifying herd size."

Previously, I noted that the Ministry of Environmental Protection's report (SOU 1995:100, p. 13) credited moderate reindeer grazing with increasing biological diversity of the pastures, while further grazing pressure reduces such diversity. In fact, the Ministry's report and its ensuing proposition to Parliament call for an alteration in the current Reindeer Herding Act's sixty-fifth paragraph so that the environmental goals of sustainability and biological diversity are to be combined with good profitability—as if the flexibility existed to make this a painless or even possible marriage. An important question therefore arises if TAQs are to be set according to the diversifying herd size as suggested: How does the diversifying herd size compare to the rational herd size? It is highly probable that the diversifying herd size, exerting moderate grazing pressure and conducive to biological variety, is considerably lower than the rational herd size, that which would supposedly secure for herders the best income. In introducing the principle, it is not my intention necessarily to recommend it as the optimal criterion for herd sizing, but simply to identify it as one of the implicit factors which drive the negotiations on "preserving nature."

The difficult task of trying to assess an actual rational herd size for any Sameby, involving as it does negotiation of all the variables we have so far considered (of course within their somatic bounds) becomes all the more problematic when forms of exploitation external to the regular herding system, such as logging practices, and other changes in the landscape are considered. While the practically defined TAQ is not identical to the conceptual rational herd size, it is generally set at a level quite sufficient to protect the grazing resource from over exploitation. Most importantly, it is the TAQ which by decree of the herding authorities is to regulate actual reindeer numbers.

The reindeer statistics from the herding authorities indicate that, at least since 1920, total reindeer numbers for Sweden as a whole first came to exceed its TAQ total of 280,000 head in 1986. The most recently available figures (from the 1993/94 count) show that the high reindeer numbers relative to TAQs can be localized mainly to the Västerbotten county's mountain Samebys and the Jämtland county's southern Samebys. According to the Board of Agriculture's reindeer herding section, successful measures had already been taken to bring most of the effected Samebys in line with their TAQs before the overgrazing debate gained momentum. (Of course deterioration of pastures, should there have been any, can take many years to overcome.)

Sticks and (Poisoned) Carrots

Centralized authority, resource users encompassed by higher-order unity and regulation, provides the means to implement negative feedback. In fact, it is the acceptance of such regulation by its constituent parts that brings a higher-order systemic entity into being or to certain extent defines it. The goal of higher-order controls and leveling mechanisms is to prevent individualized skill or maximization efforts (in excess, greed) from producing uncontrolled individual gain and consumption. Such measures strive to eliminate the aspect of the dilemma involved in commons depletion, whereby the combined effect of many individual consumers (even if each exists at a near minimal subsistence level) will destroy the resource, and whereby any individual self-restraint exercised by one consumer will only play into the hands of another without ecological benefit.

Even if the benefits to individuals of maximizing their herds cannot be entirely removed, herd reduction might be made individually meaningful by measures such as herding fees or grazing fees exacted per reindeer. The imposition of a total ceiling limit, while it might save the commons, does not eliminate competition among members of the commons which can be expressed in subtle forms and constitute the aspects of the commons dilemma in a broad sense.[8] Similarly, if one fails to impose maximal quotas for individuals before a total ceiling limit has been reached, one does not eliminate competition among commons members and dilemma until that ceiling is reached and stabilized *with all commons members at their component maximal quota levels.* If, for example, the Sameby's reindeer limit is exceeded and some herders are far above their individual limits while others are far below, once those who have too many decrease their stock to bring the Sameby total under its collective limit, there will be a scramble among the other herders to fill any available room with their *own* deer. In fact, should the smaller herders continue to increase their holdings toward their individual limits and thereby push the Sameby total over the top again, it is not they who will be forced to cut back, but those who are still at that time over their individual limits. In short, those under their individual limits have no reason to cut back even with the total Sameby reindeer population at or over its limit.[9]

In a pastoral system, it is only when individual quotas are enforced—*whether or not the total ceiling limit which encompasses their sum has been attained*—that competition among commons members with regard to grazing is avoided. Note, however, that such a system can be distinctly at odds with rationalization policies and the stated goals of "sustainable development" which seek to utilize available resources sustainably *yet fully.* Should an industrious herder reach his individual quota long before the Sameby's total ceiling TAQ is achieved, restraint of his herd growth appears wasteful.

One might well argue that in the modern world, where reindeer herding is so strongly dependent upon the market economy and the consumption of reindeer meat by non-Saami, where other land-using industries compete with the reindeer, and where local environments have become global issues, traditional Saami pastoral mediation and ecological regulation is no longer sufficient, not even for successful herding or for the best interests of the Saami. While a strong case can be made for the necessity of centralized authority on one level or another, it is important to note that this need not imply Swedish authority. There is the cooperation and sharing of traditional Saami herding groups; there is the Sameby unit (even if it is largely a Swedish construct as it is organized today); there is also now, since 1993, the Saami Parliament (Sameting). The difficult decisions regarding resource access and utilization levels among the Saami would not disappear just by being transferred from state to Saami authority. The essential point, however, is that they should be Saami decisions. Only then can Saami skills maintain continuity while keeping pace with new developments.

This is not to say that the content of regulatory mechanisms would necessarily be the same or that the vital point is simply that the same mechanisms should be proclaimed and implemented by the Saami rather than by the state. Admittedly, the source of the authority is important in itself, but this can have implications as well for the specific approach implemented in those decisions. Faced with the same regulatory problems, the Saami may give quite different regulatory responses. Should the Saami refrain from instituting a centralized Sameby herding authority or individual herd limits, this would not be a condition of no regulation, but rather regulation left to other devices such as inter-Saami competition.

Saami herders are not in the least opposed to modernization, if with Paine (1994:141) one understands this term to cover pastoral life changes which have occurred without being "part of the concerted state program" (this being rationalization), and much of pastoral adjustment and improvement stems from herder competition within a system of freely taken risks. However, state impositions are geared toward suppression of herder competition (once the small herders have been removed and the "proper" components of the state model established). Regulation of the commons dilemma has been approached precisely by establishing mechanisms designed to prevent individual efforts or skills from producing unlimited individual gains. But how can one withdraw the benefit of individual skill without in time severely diminishing skills?[10]

With reference to the fact that some Swedish farmers are subsidized to keep the livestock necessary to preserve meadows and enclosed pastures by their grazing, Margareta Ihse suggests in her debate article that the "herding Saami [should] also obtain an economic acknowledgment for their formation and protection of our unique natural and cultural heritage in the mountains" (in *Samefolket* 1995:15). This appears to be a diplomatic way of suggesting that the Swedish state should pay Saami herders for their contributions to the care and maintenance of the mountain regions. If such a policy were put in place, then possibly those Saami who might otherwise be economically ruined by the reindeer cut-backs necessary to avoid destruction of the mountain environment could continue their herding livelihood with moderate grazing pressure. This could be good for vegetative diversity as well as for the "natural norm" in the mountains considered desirable by the Swedish (and maybe even international) public. Another, more threatening method of linking subsidies to environmental goals, and that suggested by the Ministry of Environmental Protection (SOU 1995:100, p. 40), is that already existing subsidies be withheld from herders in Samebys that exceed their TAQs in order to achieve herd reductions. According to this vision, the Saami herding livelihood, along with the culture and traditional skills it maintains, can best be integrated with environmentalist demands by government subsidies.

Anders Sirén, however, rejects Ihse's suggested subsidy:

> Reindeer herding has the strength to survive in one form or another, and any kind of special subsidy for "formation and protection" of "natural and cultural heritage" is unnecessary and unjustified. Clear and appropriate playing-rules are all that reindeer herding needs. It would also help if one could solve separately the question of the status of Saami in Sweden so as to be spared from the constant confusion between the Saami as a people and reindeer herding as an industry. (Sirén, in *Samefolket* 1995:16)

Sirén has asserted that the reindeer livelihood has the strength to survive without special financial compensation for environmental stewardship according to the Swedish "natural norm." Those who consider reindeer herding to be viable economically, socially and culturally (as do I) do not like to see new government crutches provided for its maintenance should they then open the livelihood to attack by the majority as being a costly, economically pointless hobby for a few "culture-carriers" at general expense. Yet this is a battle which can hardly be avoided with or without the creation of an environmental protection subsidy. Reindeer herding already enjoys a number of government subsidies and funds for protection against disaster. In any case, the degree of monetary support must be deemed insufficient for one to reach a meaningful evaluation.

Reindeer meat competes with other meat products. The reindeer industry competes for scarce resources with other land-based industries, and the number of people employed by the reindeer industry is continually ranked against the number of people employed by non-herding and competing land-based industries when it comes to counting social benefit. Reindeer herding is indeed subsidized, but so are other competing forms of employment and meat products. There can be no meaningful evaluation of the independent strength or weakness of the herding industry, its ability to do without subsidies or its need for them, for such matters can only be considered in relative terms, that is, in comparison with the subsidies and special benefits, legal as well as economic, accruing to the competition.

The government's own accountants for 1994/95 report fully thirty-five different forms of support and compensation to the herding industry (Riksdagens Revisorer, Rapport 1995/96:8). In fact the same report asserts that the reindeer-herding industry receives in subsidies eighty percent of the value of its final production (Riksdagens Revisorer, Rapport 1995/96:8, p. 72). Should forms of compensation payments, for example, payments for reindeer losses occasioned by protected predator species, be added to the subsidies, the accountants calculate the degree of support to be 178 percent of production value. However, other industries also receive much by way of support. (Here one must also count tariff protection and tax breaks.) Of course, one can question the manner in which certain funds have been counted. If the state decrees that certain reindeer predators are protect by law, prohibiting the hunting of them and preferring instead to pay herders for the reindeer losses these predators occasion, should such funds be counted against herding profits? If rangelands are destroyed by the construction of a huge hydroelectric power dam and the Samebys impacted are compensated in a one-time monetary payment for their loss of grazing, is this to be counted as government *support* to the herding industry? The accountants acknowledge such arguments, but conclude that however one counts, "the reindeer industry enjoys support which is not insignificant in relation to its production value" (Riksdagens Revisorer, Rapport 1995/96:8, p. 72).

In the past all reports decrying the economic situation of the reindeer industry have led without fail to the conclusion that herd management must be further rationalized. Since the meat production of the reindeer industry is so insignificant on a national scale—now counted at a few tenths of a percent of all the meat produced in Sweden (Riksdagens Revisorer, Rapport 1995/96:8, p. 73)—the essential reason for advocating such rationalization has been to improve the income and living standards of the herding families (welfare ideology). The report of the government accountants, however, departs from a sole concern with welfare ideology and instead swings toward the new environmentalist perspective.

> The reindeer industry has values of entirely different nature for Sweden than the pure socio-economical. With this insight as a foundation, it should be possible to steer away from thoughts of strict rationalization and effectivization which anyway have small chances of success and which can besides come into conflict with other goals of a superior type, for example, environmental goals. One must not forget in this context that a highly mechanized reindeer herding means stress on a nature which all wish to protect and increases the risk for conflict with other users of the Saami settlement area. (Riksdagens Revisorer, Rapport 1995/96:8, p. 73)

This is not to say that the report of the government accountants recants any desire to maintain decent living standards for herding families because of superior environmental goals, only that it lays environmental, rather than just welfare, concerns as a justification for the not-insignificant aid. From the government perspective, old crutches can be given new names or repackaged. Even further funding might be called for in order to support herding under the new environmental banner, and certainly one must expect that the government will demand linkage between the payment of such aid and the ability of the Samebys to realize newly legislated environmental goals. Herders are naturally wary that the realization of Sameby environmental goals might also become linked to even other subsidy forms. Linkage of this sort, considering the current magnitude of subsidies, would undoubtedly far outweigh the incentive provided by the threat of fines for Samebys to maintain their TAQs.

Minority Rights and the Herding "Industry"

I shall refrain here from delving into the importance of reindeer herding for the Saami culture. I have discussed it elsewhere as have numerous others (Beach 1981; Ruong 1982). It should be noted however, that Saami hunting and fishing have been legally defined by Swedish law as appendices to the herding practice only, and no longer characterize in themselves a unique Saami culture or way of life. The Saami language has been classed among those found to be in an advanced process of language change-over—to the languages spoken by the majority in each of the four nations hosting an indigenous Saami population (SOU 1990:84 p. 161). Lack of employment opportunities in the north has forced many Saami to relocate to urban centers in the south. Reindeer herding both as current practice and as historical heritage supplies both herding and non-herding Saami with one of their most powerful markers of ethnic identity.

What I wish to highlight in this context is the flip side of herding/culture symbiosis: the importance of Saami culture and special Saami minority rights for the herding industry. Sirén states the belief that the herding industry would be helped if questions related to the status of the Saami in Sweden were not constantly confused with matters concerning reindeer herding as an industry. I believe this separability to be a dubious proposition. If ever attained, I believe, contrary to Sirén, that the separation of herding from Saami indigenous affairs would prove disastrous for herding not only as a livelihood supporting a considerable, though small, number of people (regardless of ethnicity), but also disastrous for herding as an industry with meat production only at heart.

In contemporary Swedish law, it is Saami rights considerations alone which reserve herding as a livelihood only for Saami. Certainly there are other indigenous issues which might conflict with the principles of "rational reindeer management," but this formulation is meaningless unless one considers for whom it is intended to be rational. What is rational for the market may not be rational for the Saami as a people. Conversely, it is true that what is best from the aspect of Saami rights and ethnic mobilization might not always make the most sense for the economic profitability of herding for those Saami engaged in herding. Be this as it may, however, the two mutually support each other. Should the Saami monopoly on reindeer herding be revoked, its moral justification would become leveled to that of any other industry making use of land resources in the north. This tendency is already clearly visible even within the context of Saami monopolization as herding becomes pushed toward a full ranching system. The more herding comes to resemble other Swedish animal farming, the more it loses in the eyes of the majority its moral justification in claiming special resource rights (according to the Swedish paradigm of privilege based on cultural preservation).

It might be argued that herding would do well, maybe even better in financial terms, were it to compete on an equal basis with other forms of land utilization, if by so doing it would be free to follow the dictates of Swedish rationalization policy and environmental ideals. But one need only glance at the historical record to realize that herding has time after time been curbed to permit the expansion of other interests—indigenous rights not withstanding. Evaluated merely in terms of how much money it brings and how many people it employs, and comparing these numbers to how much land area it needs free from forms of heavy extractive exploitation, the herding industry can hardly maintain a strong position with respect to other industries competing for land utilization. Those regional planning authorities concerned with the holistic rationalization of all forms of livelihood together—not to mention the majority of non-herding voters—cannot fail but to continue supporting other forms of utilization at the expense of herding. Herding will be tolerated to the extent that it can survive within a pattern of land utilization which holds itself to areas unattractive to its competition. That which can and should bolster the herding position is precisely its status as a Saami livelihood, invoking the moral obligation to support indigenous peoples and cultures as recognized in international conventions.

Conclusion

The day may come when the environmentalist block is so powerful in Sweden that the protection it can offer and actually secure for the herding livelihood is enough for the Saami to find benefits greater than costs in allying themselves with Swedish environmentalist ideals and regulatory enforcement. Or, should Saami livelihoods be regulated by state policies to the brink of extinction and offer no real conflict with Swedish environmental goals and exploitation, we might come to see half-mythical and overly noble Saami figures trotted forth by the Green lobby as symbols for the environmentalist movement on national television in a manner similar to that which has occurred with Native peoples in America. Currently, however, this is far from the case, and Saami see fit to fight for their own environmental empowerment and the ability to attain skills according to their own interaction with the land and their own goals. Being the masters of a landscape with a continuity rooted in their heritage is to them more meaningful than being state-subsidized employees working a Swedish model of grazing units for maximal yield.

As Ingold has clarified, persons and environments are engaged in mutual constitution (1992:40). Hence, to place the Saami herding environment within a Swedish ecological framework of regulation is to put the essence of what it is to be Saami in Swedish hands.

> Enfolded within persons are the histories of their environmental relations; enfolded within the environment are the histories of the activities of persons. Thus, to sever the links that bind any people to their environment is to cut them off from the historical past that has made them who they are. Yet this is precisely what orthodox culture theory has done, in giving recognition to the historical quality of human works only by attributing them to projects of cultural construction opposed to, and merely superimposed upon, an ahistorical nature. (Ingold, 1992:51)

While the Saami might claim superior environmental knowledge or argue that the proposed state regulatory models have essential flaws which cause them to be less effective than their own, these objections (even if true) are formulated so as to fit Swedish terms of debate. In my opinion, such flaws cannot alone account for the vehemence with which the Saami reject the Swedish program for the "environmental adaptation of reindeer herding." Intuitively, the Saami see this program as a threat to their persons as Saami, to the heritage of their ancestors, and to the future of Saami society.

---

[1] For an interesting comparative case from the Scottish Highlands, see Toogood, 1995.

[2] The thought that development, which to many is predicated upon various forms of exploitation, be it of the environment, of southern nations, or of a working class, should be sustained at all evokes a flurry of ethical conundrums. Are we speaking of development for a given, limited population, or development for one that is exponentially increasing without restraint? And even if by development we try to restrict ourselves to the attainment of basic, minimal humanitarian goals, like decent health and living standard for all, how can one possibly prevent these, when coupled to an unrestricted growth of beneficiaries, from generating an unsustainable situation exploitive of limited resources? Given these considerations, (unless it addresses the issue of human population control) the thought that indigenous knowledge—or for that matter any knowledge directed toward resource utilization—might promote sustainable development can be credited only under a term so short as to mock the concept of sustainability or in a systemically closed context hardly experienced in modern times.

[3] One can also speak of the goal ranges for the reindeer within a more inclusive system encompassing the reindeer, grazing and herders, "goal herding ranges." Human mediation makes a difference in the number of reindeer which can be accommodated sustainably on the same pasturage. Obviously it might make a major difference should this be Saami herding mediation (protected by special indigenous legislation and rooted in a continuity of Saami traditional skills) or non-Saami mediation. Recognition of this wider system and negotiation of its parameters are, of course, the essential issues of this paper.

[4] Except for the four northernmost mountain Samebys, Könkemä, Lainiovuoma, Saarivuoma and Talma, which share one large TAQ in common.

[5] Elsewhere (Beach, 1997 forthcoming) and building upon the work of Gregory Bateson, I argue that the commons dilemma is a natural part of the evolutionary process which is also a process of mind. The challenge posed by the problem of the commons is essentially a learning challenge.

[6] Instead of "strategies" (which to some might carry a connotation of rigidity), it might be better to view them as initial orientations which are then tempered, reformulated or abandoned as practicalities demand.

[7] For example, aspects of commons dilemma can surface in the realm of labor committed for the common good. Elsewhere (Beach, 1997 forthcoming) I have addressed the issue of so-called "free-riding" among Sameby members, that is, when individual herders, relying upon the labor of other herders, shirk their collective Sameby work responsibilities to invest their time in individually more profitable channels (sometimes even while drawing a Sameby salary).

[8] Should this situation be accompanied by years of good growth for the reindeer population, it will naturally result in the leveling of all individual herd sizes to their prescribed limits, and this condition would continue until reindeer losses (by slaughter, predation or the catastrophic climatic blockage of grazing access) brought about a total

Sameby herd size well under its TAQ, whereupon individual herd sizes could again begin to diverge dramatically.

[9] It seems that the state model first promotes ruthless competition among herders so that small herders will be ousted and the "proper" rational number of herders and reindeer will be established. Then inter-herder competition is to cease so that the commons tragedy is avoided, but instead competition is to surface full force again at the market level as reindeer meat competes with other meats, and at the political level where the social worth of the herding livelihood (production and employment) competes against alternative forms of land use.

References

Andersson, G. 1995. Samer och renar utpeckas som miljöbovar: hotar att ödelägga de svenska fjällen [Saami and reindeer are pointed out as environmental villains: Threaten to lay waste the Swedish mountains]. *Samefolket* (8).

Beach, H. 1980. Swedish rational herd management: Unexpressed premises and unforeseen consequences. *Proceedings of the second international reindeer/caribou symposium,* 17-21 September, 1979, Röros, Norway. Reimers, E., Gaare, E. and Skjenneberg, S. Trondheim, eds. Norway: Direktoratet for vilt og ferskvannsfisk.

———. 1981. *Reindeer-herd management in transition: The case of Tuorpon Saameby in northern Sweden.* Uppsala Studies in Cultural Anthropology 3. Stockholm, Sweden: Upsaliensis Academiae.

———. 1983. A Swedish dilemma: Saami rights and the welfare state. *Production Pastorale et Société,* bulletin de l'équipe écologie et anthropologie des sociétés pastorales. Supplément à *MSH Informations.* Publié avec le concours du CNRS. 12 (Printemps).

———. 1985. Moose poaching or Native minority right: A struggle for definition in Swedish Saamiland. *Nord Nytt, Nordisk Tidskrift for Folkelivsforskning* 26. Special-Trykkeriet Viborg a-s, Denmark.

———. 1992. Den Svenska Samepolitiken [Sweden's Saami policy]. *Invandrare och Minoriteter* 2 (April).

———. 1993. Straining at gnats and swallowing reindeer: The politics of ethnicity and environmentalism in northern Sweden. In *Green arguments and local subsistence.* G. Dahl, ed. Stockholm: Stockholm Studies in Social Anthropology.

———. 1994, Shots heard round the world. In *Beslutet om Småviltjakten - En studie i myndighetsutövning [The decision about small game hunting - a study in the exercise of governmental authority*]. Agneta Arnesson-Westerdahl, ed. utgiven av Sametinget.

———.1997. Reindeer-pastoralism politics in Sweden: Negotiating a common property regime. Forthcoming in *Practical science and scientific practice: Situated resource use in Nordic environments.* Gisli Pálsson and Torben Vestergaard, eds.

Constenius, T. and Danell, Ö. 1995. *Metodutveckling i fråga om renbetesinventeringar* [*Methodological development in the question of reindeer grazing land inventories*]. PM 1995-11-20. Jönköping, Sweden: Jordbruk-sverket.

Dagens Nyheter. 1995. Renbete förstör fjällen [Grazing reindeer destroy the mountains]. July 21.

Eide, A. 1985. Indigenous populations and human rights: The United Nations efforts at Midway. In *Native power: The quest for autonomy and nationhood of indigenous peoples.* J. Brøsted et al. eds. Oslo: Universitetsforlaget.

Ihse, M. 1995. Renskötseln hot mot fjällnaturen: samerna kan inte fortsätta skylla allt på turisterna [Reindeer herding, threat against the mountain environment: The Saami cannot continue to blame everything on the tourists]. *Samefolket* (8).

Ingold, T. 1976. The Skolt Lapps today. Cambridge: Cambridge University Press.

Ingold, T. 1992. Culture and the perception of the environment. In *Bush base: Forest farm: Culture, environment and development.* Elisabeth Croll and David Parkin, eds. London: Routledge.

Manker, E. 1947. *Det Svenska Fjällapparna* [*The Swedish mountain Lapps*]. Stockholm: Svenska Turistföreningens Förlag.

Norrbotten's Kuriren. 1994. Tramp av. renhjordar förstör fjällen [Tramping of reindeer herds ruins the mountains]. December 9

Paine, R. 1994. *Herds of the tundra: A portrait of Saami pastoralism.* Smithsonian Institution Press, for *Anthropologica* (36):232-234.

Proposition 1992/93:32. *Samerna och samisk kultur m.m.* [*The Saami and Saami culture, etc.*] 1 Riksdagen 1992/93. 1 samling. Nr. 32. Sweden.

Proposition 1994/95:100 Bil. 10. Jordbruksdepartementet. 1 Riksdagen. 1 samling. Sweden

Proposition 1995/96:226. *Hållar utveckling i landets fjällområden* [*Sustainable development in the land's mountain regions*]. 1 Riksdagen. 1 samling. Sweden

Riksdagens Revisorer. Rapport 1995/96:8. *Stödet till rennäringen* [*The Aid to Reindeer Herding*]. Stockholm

Ruong, I. 1982. *Samerna I historien och nutiden* [*The Saami in history and the present*]. Stockholm: BonnierFakta.

Samefolket. 1995. ed. Olle Andersson. Nr. 8. Östersund: CEWE-Förlaget.

Sametinget, The Swedish Saami Parliament. 1994. *Beslutet om Småviltjakten - En studie i myndighetsutövning [The decision about small game hunting: A study in the exercise of governmental authority*]. ed. Agneta Arnesson-Westerdahl.

Sápmi. 1995. The Social Democratic Party's Political Program Regarding the Saami.

SFS 1971:437. *Rennäringslagen [The Reindeer Herding Act]* Stockholm.

SFS 1993:36. *Lag om ändring I rennäringslagen [The Law on changes to the Reindeer Herding Act] (1971:437).* Uffärdad den 28 januari 1993.

Sirén, A. 1995. Naturligt att renbete påverkar fjällnaturen [It is natural that grazing reindeer affect the mountain environment]. *Samefolket.* 8.

SOU 1990:84. *Språkbyte och språkbevarande.* Under-lagsrapport utgiven av samerättsutredningen [*Language change and language preservation.* A supporting report published by the Saami Rights Commission]. Presented by Kenneth Hyltenstam and Christopher Stroud. Stockholm.

SOU 1995:100. *Hållar utveckling i landets fjällområden [Sustainable development in the land's mountain regions*]. Betänkande av Miljövårdsberedningen Miljödepart-ementet [Report by the environmental protection committee. The Environmental Ministry]. Stockholm: Regeringskansliets Offsetcentral.

Sveg Case lower court verdict. 1996. Sveg Tingsrätt, Meddelad av Svegs Tingsrätt den 21 Februari 1996 i det så kallade Renbetesmålet [Communicated by Sveg's lower court 21 February 1996 in the so-called Reindeer Grazing Case]. Dom nr. DT 12, Mål nr. T 88/90, T 70/91 och T 85/95.

Toogood, M. 1995. Representing Ecology and Highland Tradition. *Area.* 29(2):102-109.

Contributors and Editors

Hugh Beach, (Ph.D. Uppsala University, 1981) teaches in the Department of Cultural Anthropology, Uppsala University, Sweden. He conducts research for the Nordic Environmental Research Program, and the Center for Arctic Cultural Research, both located at Uppsala. He has published extensively on the Saami and reindeer herding in both his native English and Swedish.

Peter Collings is a Ph.D. candidate in anthropology at Pennsylvania State University. He has conducted fieldwork in Alaska, Tibet, and Canada's Northwest Territories, and has published several articles resulting from his work with the Inuit.

Kurt E. Engelmann, (Ph.D. University of Washington, 1993) is Assistant Director of the Russian, East European and Central Asian Studies Center (REECAS) and lecturer at the University of Washington's Jackson School of International Studies. His specialties are geography, especially Central Asian, and electronic resources.

Aileen Espiritu, a lecturer and Ph.D. candidate in the Department of History at the University of Northern British Columbia, has conducted fieldwork in Russia's Tyumen' Oblast' and in West Siberia. Her dissertation examines the impact of industrial development on the aboriginal peoples of West Siberia.

Gail A. Fondahl, (Ph.D. University of California, Berkeley, 1989) is a professor at the University of Northern British Columbia. Her research interests include First Nations' land tenure systems and land claims. She has published extensively on the indigenous peoples of Russia, and is currently undertaking a three-year project on indigenous land tenure and self-government in the Sakha Republic.

Charles Johnson serves as Executive Director of the Alaska Nanuuq Commission and a member of the International Arctic Science Committee. He has also served on the Executive Committee of the Inuit Circumpolar Conference.

Joan McCarter is a 1997 graduate of the REECAS master's degree program at the University of Washington's Jackson School of International

Studies and an author of *The post-Soviet handbook: A guide to grassroots organizations and Internet resources in the Newly Independent States.*

Eric Alden Smith, (Ph.D. Cornell University, 1980) serves as a professor in the University of Washington's Department of Anthropology. He has conducted research, published and teaches in the fields of ecology, economics, evolutionary theory, and hunter-gatherers, focusing on the Inuit in the Canadian Arctic.

Craig ZumBrunnen, (Ph.D. University of California, Berkeley, 1973) is a professor of Geography at the University of Washington. His research and publications have centered on natural resource management and conservation, methods of resource analysis, environmental quality problems, physical geography, and modeling of human impact in natural systems. He has focused particularly on the former Soviet Union.

Index